Your Essence, Your Eternity

Your Essence, Your Eternity

Copyright © 2021 by Howard Jeffrey Bender

All rights reserved. Except as permitted by U.S. Copyright Law, no part of this book may be reproduced in any form or by any electronic or mechanical means including information storage and retrieval systems without permission in writing from the publisher.

This publication is intended to provide accurate and authoritative information in regard to the subject matter covered. Reasonable efforts have been made to ensure the information is reliable, but the author and publisher cannot assume responsibility for the accuracy of all the material. Also, the author and publisher have attempted to trace the copyright holders of all material reproduced in this book and apologize to copyright holders if permission to publish has not been obtained. If we have not acknowledged copyrighted material, please contact the publisher so we may rectify it in any future reprint.

ISBN: 9798590038992
First Edition
Printed in the United States of America

Other books by Howard Jeffrey Bender

Applied String Theory

The Eternity Problem

Our Inner Alien

Gaia's Climate Challenge – Giving Humans Their Last Chance

Acknowledgements

This book is intended to help guide people to have better lives. I'd like to acknowledge four very special people who have certainly helped me do that.

Jacquelyn Green is a speech therapist I met in college because I had finally become disgusted enough with my stuttering problem that I sought help. Jackie used an unconventional approach that worked so well that I became a teacher and focused my life on becoming a better communicator. How have I done, Jackie?

Randi Anderson is an excellent mother and therapist who opens her heart whenever she deals with people and everyone lucky enough to work with her invariably benefits from that experience. There are those, including me, who sometimes take advantage of her warmth, but she always keeps her positive view on the goodness within everyone. Carry on, Randi!

Nicole Cayouette is one of those people everyone needs around them – someone with boundless energy who uses that energy to help others. A person with a big family and a bigger heart, Nicole is the one to turn to because you know she'll do whatever is possible to help. She's helped with my various but futile get-rich-quick schemes and given the most sensible advice on my incessant dreamings. Thanks, Nicole!

Dr. Susan Gartner is my essence mate, the one who makes every day a day to look forward to. Everyone likes Susan because she's so nice, which is a bit disgusting because she's also so talented. How many people do you know who can win a city tennis tournament and also get the Best Sportsmanship award? Trained as a biochemist, Susan spends her time and energy helping make the world a better place, and some of the personality improvements I suggest in this book are those I see every day in Susan.

Your Essence, Your Eternity

Dedication

This book is dedicated to the essence of

Heidi Julia Bender

who told the world from her wheelchair

"I want people to know that they can do whatever they want, no matter what their situation is. They can do it – they shouldn't give up."

Table of Contents

Acknowledgements iii
Preface x
Part 1 1
The Essence and Eternity 1
   Chapter 1 3
   Life 3
      Physical Body 7
      Habitat 10
      Culture 11
      Essence 16
      Consciousness 17
      Reincarnation 24
         Normal Explanations 27
         Paranormal Explanations 30
   Chapter 2 35
   Origin and Evolution of Life 35
      The Origin of Living Things 35
      How Life Developed 37
         Archaea 41
         Bacteria 41
         Eukaryota 42
      Building Blocks of Life 43
         Beginnings of Life 46
         Living Things 51
      Overview of Evolution 55
         Biological Evolution 58
   Chapter 3 65
   Sensing Our Habitat 65
      Reality 65
      Your Five Senses 66
      Personalizing Reality 74
      Essence Location 76
   Chapter 4 79
   Essence and the Brain 79

Brain	80
Brain Structure	82
Human Brain Structure	88
Human Brain Functions	92
Memory	96
The Mind	96
Brain Development	101
Essence in the Brain	105
Your Exclusive Essence	107
Chapter 5	111
Communication	111
Genetic Communication	111
Communication Through Learning	113
Directed Communication	115
Essence Communication	117
Historical Experiences	120
Essence Communication with the Brain	121
Chapter 6	123
Essence In String Theory	123
Background	125
String Theory Dimensions	127
The Essence Dimension	130
The Essence Process	133
Part 2	137
Improving Your Essence	137
Chapter 7 – Your Personality	139
A Healthy Personality	140
The Golden Rule In Practice	147
Chapter 8 – Your Essence Development	151
Stages of Development	152
Essence Maturation	155
Clones	159
Personality Development Factors	163
Essence Anomalies	166
Essence Age	166
Chapter 9 – Your Essence Traits	173
Personality	173

Personality vs. Behavior	177
Personality Characteristics	186
Understanding Your Personality	188
Personality Attributes	190
Essence Personality Attributes List	194
Each Personality Attribute	199
Openness	199
Liveliness	199
Abstractedness	203
Openness to Change	209
Philosophical Attitude	219
Conscientiousness	225
Perfectionism	225
Maturity	232
Achievement Attitude	240
Task Performance Attitude	249
Control Attitude	259
Vigilance	269
Dependability	279
Extroversion	283
Warmth	283
Privateness	290
Material Attitude	295
Risk Attitude	306
Leadership	320
Socialization	326
Agreeableness	339
Ethics	339
Morals	347
Sensitivity	369
Aggressiveness	380
Fairness	390
Neuroticism	398
Emotional Stability	398
Apprehension	406
Tension	410
Egocentrism	414

- Chapter 10 – Your Essence Interactions 427
- Essence Interaction with the Physical Body 429
  - Openness...431
  - Conscientiousness..................................432
  - Extraversion...434
  - Agreeableness..442
  - Neuroticism...446
- Essence Interaction with the Culture..............453
  - Openness...454
  - Conscientiousness..................................457
  - Extroversion..461
  - Agreeableness..465
  - Neuroticism...475
- Essence Interaction with the Habitat..............477
  - Openness...478
  - Extroversion..481
  - Neuroticism...489
- Chapter 11 – Your Life of Chance....................491
  - Getting the Breaks 491
  - Probabilities...498
  - Why Things Happen................................508
- Chapter 12 – Improved You 517
  - Summary...517
  - Improvement Plan 518
- Notes 523
- References 529

FIGURES

FIGURE 1-1. COMPONENTS OF LIVING THINGS 4
FIGURE 1-2. MASLOW'S HIERARCHY OF NEEDS 9
FIGURE 1-3. HUMAN CULTURE AND TECHNOLOGY 15
FIGURE 1-4. GRAVITY DISTORTION OF SPACE-TIME 22
FIGURE 2-1. TIMELINE FOR LIFE 38
FIGURE 2-2. TARDIGRADE 43
FIGURE 2-3. DNA STRUCTURE 45
FIGURE 2-4. GENERAL CLASSIFICATION OF LIVING THINGS 52
FIGURE 2-5. LIGHTER COLORED PEPPERED MOTH 59
FIGURE 2-6. DARKER COLORED PEPPERED MOTH 59
FIGURE 2-7. MUTATED PIG 61
FIGURE 3-1. THE VISIBLE SPECTRUM 67
FIGURE 3-2. CARINA NEBULA VISIBLE 68
FIGURE 3-3. CARINA NEBULA INFRARED 68
FIGURE 3-4. CHARMING A SNAKE 70
FIGURE 3-5. STARNOSED MOLE 73
FIGURE 4-1. NEMATODE C. ELEGANS 83
FIGURE 5-2. C. ELEGANS NEURON CONNECTIONS 84
FIGURE 4-3. NEURONAL COMMUNICATION 85
FIGURE 4-4. BRAINS OF VARIOUS MAMMALS 86
FIGURE 4-5. BRAIN COMPARISONS 87
FIGURE 4-6. LOBES OF THE CEREBRAL CORTEX 90
FIGURE 4-7. HUMAN BRAIN STRUCTURE 91
FIGURE 4-8. SPECIFIC HUMAN BRAIN FUNCTIONS 93
FIGURE 4-9. PHINEAS GAGE 94
FIGURE 4-12. EMOTIONAL ANALYSIS AREAS 103
FIGURE 5-1. FLATLAND HOUSE 118
FIGURE 6-1. SPATIAL DIMENSIONS 128
FIGURE 6-2. ESSENCE DIMENSION 130
FIGURE 6-3. INITIAL ESSENCE LINK 134
FIGURE 6-4. BRAIN-ESSENCE LINK THROUGHOUT LIFE 135
FIGURE 8-5. BRAIN DEAD 135
FIGURE 8-1. WILLIAM AND JAMES BULGER 158
FIGURE 8-2. IDENTICAL TWINS 158
FIGURE 8-3. CHANCE THE BULL 160
FIGURE 8-4. CLONED MICE IN A MOUSE PARADISE 161
FIGURE 10-1. PERSONALITY ATTRIBUTE INTERACTIONS 429
FIGURE 11-1. ADVERTISEMENT FOR LUCKY JEWELRY 494
FIGURE 11-2. TYPICAL NORMAL CURVE 500
FIGURE 11-3. MOST EVENTS OCCUR TO THE LEFT OF CENTER 501
FIGURE 11-4. MOST EVENTS OCCUR TO THE RIGHT OF CENTER 501
FIGURE 12-1. PERSONALITY IMPROVEMENT 520

Preface

This book contains two parts. The first part describes how your physical component (body) is separate from your essence component (reflected in your personality) and how your essence may be eternal. Well, if your essence lives forever, wouldn't you want to have the best one you can? The second part helps you do that by describing what contemporary psychological research says makes your personality the way it is; how to handle people who have poor personalities; and how to recognize and improve your personality traits that you don't like.

To unambiguously state what they're trying to do scientists often use mathematical formulas, and some of the research referenced in this book is described mathematically in the published papers. However, as Stephen Hawking notes in *A Brief History of Time*, "Someone told me that each equation I included in the book would halve the sales. I therefore resolved not to have any equations at all" (p. vi). I tried to follow that sensible advice and hope I haven't lost half my readers by mentioning Einstein's equation $E=mc^2$.

The conjectures and conclusions in this book are solely my own, except those that are specifically noted as being from somewhere else. Any errors or omissions are also mine. Finally, I want to make it clear that none of the authorities I've referenced have expressed support for any of my conjectures and conclusions.

So, if you're interested in your past, your present, and your future, turn the page and we'll start presently.

Part 1

The Essence and Eternity

Your Essence, Your Eternity

Chapter 1
Life

Life is more complicated than the physical features you see, even those you see through microscopes. Let's start by comparing life to a lifeless object – an automobile.

When looking at a car all you see are its physical body parts, which is fine if it's just sitting there. To be functional, automobiles need a place to operate, like a road or, for the rugged types, just a relatively flat space. For city driving there may also be stop signs, traffic lights, etc. I call all this the automobile's habitat. Also, each place has its rules of the road, such as which side to drive on, whether safety equipment (seat belts) are required, and so forth. I call those aspects the automobile's culture. Of course, you also need a driver. Manufacturers are moving towards the "driverless" car, using high-tech features such as the global positioning system (GPS), various cameras, and artificial intelligence computers to make the whole transportation process rather invisible to those who want that, with those non-human features added to the car's physical body. Human drivers will still be around, with their individual personalities, attitudes, morals, ethical evaluations, etc. to add that level of independence to the road. I call this the automobile's essence. Putting the car's physical body, habitat, culture, and essence together creates the transportation system we use.

I suggest that living things can be described with those same four separate components: their physical body, their essence, their habitat, and their culture. While separate, these components overlap and interact with each other, as shown symbolically in Figure 1-1.

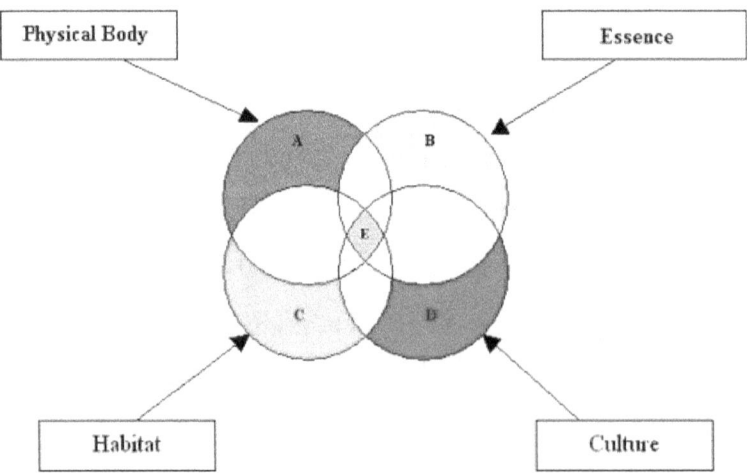

Figure 1-1. Components of Living Things

Life

In Figure 1-1, circle A represents the physical body, circle B represents the essence, circle C represents the habitat, and circle D represents the culture. The colors in these circles indicate the aspects that aren't affected by the other components, except area E as the common part of all the components together. The uncolored parts are areas of overlap, depending on the living thing represented by them. Figure 1-1 is really at least in three-dimensions, with the circles being spheres (or objects in higher dimensions). Representing these components in two dimensions makes it difficult to visualize the separate interactions.

Area E represents the influence of all the components interacting with each other and indicates the overall balance of an individual at a particular time. In Figure 1-1, the physical body, essence, habitat, and culture all have equal influences on the individual, but this is rarely the case. For instance, you may be in a culture that strongly influences how you behave, but another person may be in a weaker culture that has less influence on him/her. One individual may live in a harsh habitat that requires significant attention just to stay alive, while another individual may live in a lush habitat that demands little attention. The components are dynamic – area E is constantly changing as aspects of the physical body, essence, habitat, and culture change throughout the life of the living thing.

The interaction of these components expresses the qualities of each living thing. We humans often want to gauge the success of our qualities and conduct polls to get answers. For example, a Pew report found that married people are happier than those who are unmarried; people who worship frequently are happier than those who don't; Republicans are happier than Democrats; rich people are happier than poor people; Whites and Hispanics are happier than Blacks; and sunbelt (warmer climate) residents are happier than those who live in the rest of the country. They also found that people who have children are no happier than those who don't, after controlling for marital status; retirees are no happier than workers; and pet owners are no happier than those without pets. Do you agree? Obviously our lives involve complex interactions among the components. Let's describe these components and what it means when they interact.

Physical Body

The physical body is the component that includes all the physical parts of an individual, namely the skin, all internal organs and systems, all body fluids, and everything in the body cells. The individual's senses are also part of its physical body. All of these organs and systems function via biochemistry, that branch of chemistry that focuses on how living things operate. Biochemistry deals with proteins, lipids, carbohydrates, nucleic acids, etc., but the interaction that results in a living thing goes beyond the interactions of the biological elements. There's an additional component, consciousness, which gives the physical body the spark of life. Consciousness is not physical – it's a force of nature that differentiates the biochemistry of a rose from the chemistry of a rock. Without consciousness, physical bodies would have no life, so for our purposes consciousness is included with the physical body.

Figure 1-2 shows Maslow's Hierarchy of Needs, intended to illustrate features of living important to living things, specifically humans. These needs, in order, are:

1. Physiological
2. Security
3. Belonging

4. Esteem
5. Self-Actualization
6. Understanding
7. Esthetic Appreciation
8. Spiritual

Levels 1-5 are the basic needs while the top three are addressed after the basic needs are met. The first three describe needs of the physical body; the fourth (Esteem) describes needs for both physical and psychological health, and the fifth and above describe psychological needs. As an individual moves up the hierarchy, needs at lower levels are no longer prioritized. Deficiencies must be met first and then personal growth needs are addressed. For example, a successful businessman whose wife has just run off with a banjo player must address his "love and belonging needs" before he can resume his esteem in business growth.

Life

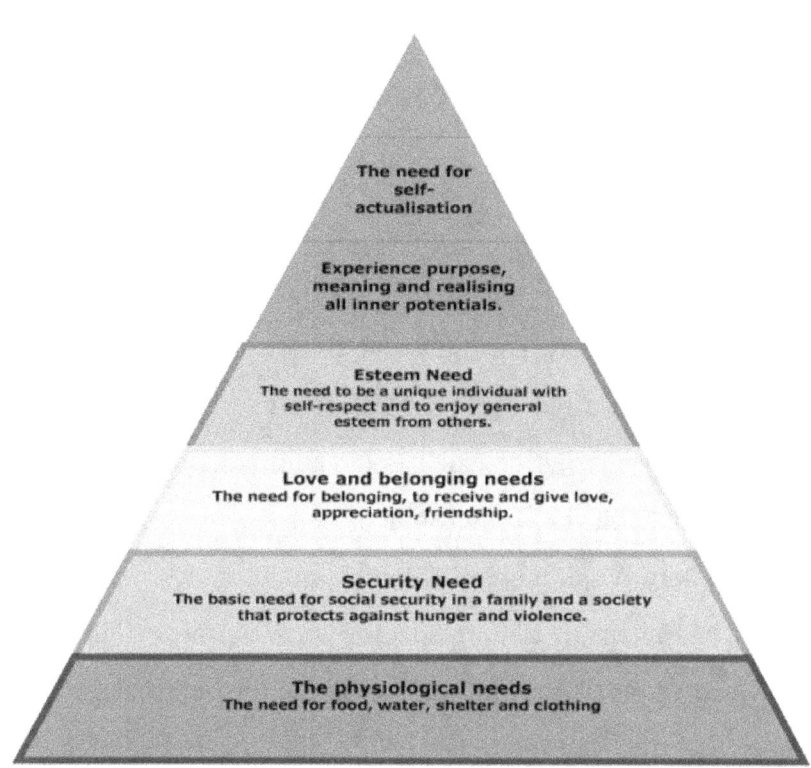

Figure 1-2. Maslow's Hierarchy of Needs

We celebrate physical bodies when they're born and mourn them when they die. Most brain functions (including cognitive thinking, sensations, and dreaming) are part of the physical body. When a new living thing is born, the only aspects that are passed from one generation to the next are those associated with the genetic code of the physical body.

Habitat

Each living thing has a habitat, which consists of the environment selected (through choice or chance) in which the living thing lives. The habitat includes the physical attributes of temperature, terrain, altitude, etc. and also the living attribute of availability of predators/prey. Habitat has been shown to affect the physical characteristics of living things through evolution, such as the ability of some animals to blend in with their environment and the loss of sight of some underground animals. Habitat isolation has permitted the development of unique species (such as those found in the Galapagos Islands) and has permitted the continuation of species (such as the lemurs of Madagascar because of lack of predators, while lemurs in neighboring Africa have become extinct).

Habitat characteristics affect the physical body by providing more or less food (resulting in larger or smaller living things), by the temperature range (resulting in more or less body fat); by the altitude (resulting in more or less lung capacity); by the availability of rain (resulting in novel ways to store water, such as a camel's humps), etc. Habitat affects the essence by providing an environment comfortable enough for less selfish characteristics to develop or by providing a harsh environment that encourages selfish characteristics. And habitat affects culture by providing space and resources. Too little space or too few resources put pressure on a culture, perhaps making it more restrictive, while a comfortable amount of space or resources may permit more cultural freedom.

Culture

Basically, a culture is the shared language, traditions, or beliefs that set a group of individuals apart from others. Cultures often develop special rules or behaviors their members must follow or particular abilities they're expected to have. For example, the Aboriginal culture expects its members to be able to survive in wilderness areas that would likely cause a New Yorker to starve to death. In the same way, a person from an Aboriginal culture would be dangerously out-of-place in New York City. Culture is something each living thing learns as it lives.

Your Essence, Your Eternity

Culture is the way that groups of people transfer information about themselves across generations and also the way dispersed people can maintain similar beliefs and activities without ever seeing each other. It's the set of shared attitudes, values, goals, and practices that characterizes an institution, organization or group. Culture includes the rules, mores, and expectations of the society in which the living thing lives. For example, ants and bees have highly structured societies with individuals physically designed for special functions (food gathering, fighting, breeding, etc.). Such cultures would have a larger influence on the physical body component than for individuals in another culture such as penguins, consisting of small family groups in colonies and who all have the same physical body structure.

All cultures change over time, with the addition of new traits and the removal of old ones, depending on changes in the other life components (physical body, essence, and habitat). Some cultures change more quickly than others and we shouldn't forget the many sub-cultures that change at different rates. For example, Americans seem to embrace new gadgets and technologies yet cling to the English system of measurement rather than converting to the metric system, because the American culture is very resistant to that change. Some cultures won't permit increased economic and political opportunities for women because of the likely changes they would bring to their culture, and it's not just the men who don't want those changes.

While we're at it, some aspects of culture are gender-specific with some traits, such as clothing and body language, suitable for males and others suitable for females. People from other cultures sometimes misunderstand what's being represented by these traits. Proper understanding of gender-specific expectations helps to clarify relationships and makes individuals better mothers and fathers. In addition, males and females sometimes react differently when faced with the same cultural situation. These differences were the theme of John Gray's book *Men are from Mars, Women are from Venus*, in which the different ways men and women react to the same situation are described. While an appreciation of such differences can help men and women better understand each other, the usual reaction is that both males and females see the other gender as acting inappropriately. At least in that way they're the same.

It's quite common for individuals to project their culture onto others living in other cultures, naturally presuming that their own culture is so superior that everyone everywhere will want to embrace it. To their surprise, this usually isn't the case and the presumptive individual is ignored (or worse). This is sometimes called *ethnocentrism* and is frequently an aspect of prejudice – you're shocked by the "alien" ways of another culture and broaden that feeling to dislike everyone in it. For example, dogs are favored pets in Western cultures but are considered to be dirty animals in Muslim cultures and are more likely to be kicked than petted. In some Southeast Asian cultures dogs are also a source of food for humans. Different cultures have their own perspectives on many aspects of life.

Figure 1-3 is a graph of the development of human culture and technology over time. The figure shows that as human societies became more complex the technology needed for the people to survive increases. Efficient agriculture needs plows; efficient cities need transportation and utilities, etc. For human culture, it appears, technology is an essential ingredient.

Life

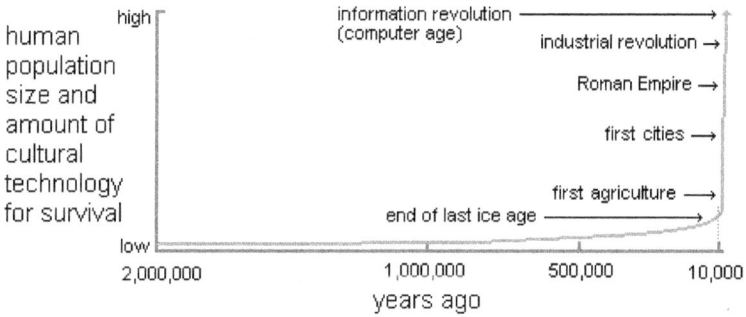

Figure 1-3. Human Culture and Technology
From Dennis O'Neil at Palomar College

Technology, however, is a two-edged sword in culture. When changes come too quickly individuals feel displaced and unsure of their place in society. The duration of company employment (lifetime), the developments of major new technologies (once or twice a century), and even the way human relationships are built (face-to-face) have all been completely changed in recent years. Alvin Toffler coined the phrase *future shock* to describe this situation, and he was writing before personal computers, the Internet, or cell phones were invented. The pace of cultural change has become so blindingly fast that even the people developing the technologies causing it can't keep up. Have we lost control of our culture?

Essence

The essence is the component dealing with the non-physical aspects of living things, including their personality, attitudes, morals, ethical evaluations, etc., and is associated with the individual's brain. The term *essence* is used to separate this component from any religious terms. Behaviors are physical body attributes learned by individuals through their cultures and habitats and are not part of their essence, although they strongly affect each other.

Newberg, D'Aquill, & Rause point out difficulties in separating mind from body:

> Neurology cannot completely explain how such a thing can happen – how a nonmaterial mind can rise from mere biological functions; how the flesh and blood machinery of the brain can suddenly become "aware." Science and philosophy, in fact, have struggled with this question for centuries, but no definitive answers have been found, and none is clearly on the horizon. (p. 32)

Anne Tyler describes the separation of physical body and essence in her novel *The Accidental Tourist*:

> Odd how clear it suddenly became, once a person had died, that the body was the very least of him. This was simply an untenanted shell ... (p. 306)

Each highly evolved living thing links to an essence, but the expression of its capabilities is determined by the type of its life form "container" (physical body), modified by its habitat and culture. The collection of these capabilities is the essence. Note that intelligence is an aspect of the physical body, not the essence. That means that human cognition, including awareness of the passage of time and curiosity about their essence, are part of the physical body, not the essence.

Consciousness

Before we go further into the essence, we should clarify how consciousness differs from it.

The scientific method has such trouble dealing with unseen phenomena like consciousness and the essence that they're considered more subjects for philosophy. Unlike the physical sciences where a theory can be quickly discarded as experiments are repeated, philosophies are more a person's opinion. Nobody can say for certain who's right – it's just an opinion. And everyone has an opinion.

If you investigate the various explanations of consciousness, some of them will seem totally stupid to you. OK, that's your opinion. To avoid getting overwhelmed with thousands of philosophies on consciousness, we'll focus on those that have been accepted by people who have made philosophical thought their careers.

Different philosophers have located consciousness in various places, from the pineal gland inside the brain to the *Res Cogitans*, a non-physical place without extension that Descartes identified as the soul. Some say consciousness exists in everything in the universe (stars, rocks, light, people) while others claim it doesn't exist at all. There's also a group of scientists who say that classical definitions of matter and energy are inadequate to describe consciousness and that you have to go to the quantum level to understand it. The theories they propose for the *quantum mind* are quite controversial.

In general, philosophers divide consciousness into *phenomenal consciousness* and *access consciousness*. Phenomenal consciousness is the state of being conscious, such as saying "I am conscious." It's related to the quality of your consciousness, such as the taste of fine wine (or bad wine), the pain of a headache, or the array of stars above you at night. Access consciousness is being conscious of something in particular, such as saying "I am conscious of my dwindling financial situation." Access consciousness includes awareness, conscience, and your intentions. Religion involves both types of consciousness while access consciousness is innately part of essence personality attributes.

Only living things are conscious, irrespective of how much you think your computer is out to get you. However, the description of what it means to be conscious has stymied science and has (so far) been relegated to explanations that sometimes seem rather mystical. In fact, Daniel Dennett points out that there are many people who hope that consciousness will always be mysterious.

An essence can be in the brain by one of two mechanisms. Either the brain creates it through biochemical means (the explanation favored by virtually all brain researchers) or the brain receives an essence transmitted from elsewhere. The "receiver" concept was first suggested in the 1890s by psychologist William James. A similar mechanism is used in television. When a television set is turned on, it receives signals existing in the air; when that set is turned off, the signals in the air still exist but are not evident by people staring at the blank screen.

Even though we know little about consciousness we often adjust it, as anyone who has drunk too much fine wine (or bad wine) can attest. Mind-altering drugs, both vilified and praised depending on your philosophy, affect both consciousness and the essence. Anesthesiologists test the level of the patient's consciousness by observing their alertness and responsiveness.

Most of the descriptions about and research on consciousness and personality refer to highly evolved living things such as animals, and some seem restricted to capabilities found only in humans. But who can deny that a plant has a level of consciousness by its defenses against predators?

Consciousness is a force of nature, like gravity, but its effects are seen only in living things. I contend that it's part of every living cell of every living thing. Consciousness is the single criterion that determines if a thing is a living thing, and includes awareness of the surroundings, decision making, fight/flight evaluations, and cooperation with other living things, as appropriate for the cell. Derek Abbott et al.

suggests that quantum mechanics plays a non-trivial role in biology. The key to life remains to be found.

Scientists have avoided describing consciousness, preferring to define a thing as a living thing through its activities or its physical nature. For example, you might say a thing is living if it is created, consumes food, generates offspring, creates waste products, and eventually dies. But fire has those characteristics, too. To be a living thing, fire would need the mysterious consciousness spark.

Perhaps an easy way to understand both consciousness and the essence is to view them in the same way as gravity. What is gravity? It's the mutual attraction of two objects that have masses. You stay on the ground because your body mass is attracted to the Earth mass, and the other way around. The Earth stays in orbit around the sun because the Earth is attracted to the mass of the sun and, by the way, the sun is attracted to the mass of the Earth. Einstein, in his Theory of General Relativity, suggested that gravity is a distortion in the shape of space-time (Figure 1-4) and should be considered a fourth dimension.

Figure 1-4. Gravity Distortion of Space-Time

But why should objects with mass be attracted to each other? Scientists have yet to find a force or a particle or *anything* that would require such attraction. But we certainly know gravity exists.

While gravity is associated with every mass, consciousness and an essence are only associated with living things. Consciousness is the main driver of the biochemistry of every cell of every living thing. When the biochemistry stops, the cell dies and consciousness disappears. Consequently, for our purposes, consciousness is considered part of the physical body.

As the most basic force associated with life, consciousness should be the easiest to identify and isolate. Unfortunately, the only real result has been the polarization of people who say it's fiction and those who say it's fact. In the 1930s Wilhelm Reich, an influential psychiatrist who worked with Sigmund Freud, said he had discovered a primordial cosmic energy he named *orgone* that was a universal life force, and he built "orgone energy accumulators" to use with his patients. The U.S. Food and Drug Administration (FDA) said the claimed health benefits were false and obtained an injunction against the interstate sale of orgone accumulators, which Reich ignored. The FDA burned several tons of his publications, arguably one of the worst examples of censorship in U.S. history, and Reich was sent to prison, where he died a year later.

More of that understanding may be at hand. In 2010 the J. Craig Venter Institute announced that they removed the DNA from a bacterium and substituted synthetic DNA they had made. Sure enough, the bacterium changed its activities to conform to the new genes in the synthetic DNA. This is still a long way from "creating life", but it's certainly a start. When scientists can artificially create a cell that does everything an amoeba does, they can claim artificial life has been created. Then we'll be more confident that, over the long history of evolution, such chemicals found their way together to start the original life and consciousness will no longer be as mysterious as it is now.

We'll also see that while physical bodies are mortal, essences are not. Suggesting that our essence is immortal sounds like reincarnation, which has to have supporting peer-reviewed evidence or it's nonsense. As it happens, data collected by University of Virginia researcher Ian Stevenson is that support. Stevenson's methodology and results have held their own against the army of skeptics who refute such claims. In addition, Dr. Lester S. King, Book Review Editor of the *Journal of the American Medical Association*, wrote that "in regard to reincarnation [Stevenson] has painstakingly and unemotionally collected a detailed series of cases from India, cases in which the evidence is difficult to explain on any other grounds.... He has placed on record a large amount of data that cannot be ignored." Let's see what Stevenson's talking about.

Reincarnation

Reincarnation means different things to different people. For example, Bill Bryson points out how the atoms in your body used to be in stars and millions of previous organisms, probably even famous individuals such as Shakespeare (or Jack the Ripper). He says:

> So we are all reincarnations – short-lived ones. When we die our atoms will disassemble and move off to find new uses elsewhere – as part of a leaf or other human being or drop of dew (p. 134).

But Stevenson's talking about the non-physical side of reincarnation rather than the physical side. What can be said about that?

Stevenson and his colleagues investigate reports of children who, just after they learned to talk between the ages of 2-4, remembered details of a past life. Most of those remembrances, while stunning in their detail, are usually relatively short snippets. The story of Kemal Atasoy, a 6-year-old Turkish boy, is especially interesting. Dr. Jürgen Keil, a psychologist from Australia, listened as Kemal said that he had lived in Istanbul (which was 500 miles away from his home) and had been a rich Armenian Christian named Karakas who lived in a large three-story house. Kemal's parents were Muslims who didn't know any Armenians and no one in Istanbul, and that part of the city had no Armenians. After several fruitless trips, Dr. Keil found a well-respected local historian who confirmed that a rich Armenian had lived in that house but died around 1940, and his family now lived in another part of the city. The historian also confirmed other details of Kemal's description, including bags that Karakas carried and his Greek Orthodox wife. How had Kemal been able to describe in such detail a house he had never seen and a man who died 50 years before he was born, whose information was so hard to find by even an experienced researcher?

Stevenson's work suggests that a person can learn from previous life experiences and use that information in future lives. This implies that our current life includes memories, behaviors, and emotions from previous lives. However, Stevenson also found that connections with a previous life disappear in the new physical body over time, with memories fading by the time the child is eight years old. A happy effect is that reincarnated children continue to express love for their previous family. Love endures.

Reincarnation is a paranormal answer to the normal question, "What happens after I die?" Perhaps the individuals claiming reincarnation as the answer have ignored other options. Jim Tucker, a researcher with Stevenson, outlines other explanations for experiences claimed from reincarnation:

Normal Explanations

Fraud. In this case, the individuals claiming to be reincarnated and their families are intentionally lying to the researcher. What motive would they have to do that? There's no monetary reward for such claims and some children have described past lives distinctly inferior to their current life. In societies such as India, from which many reincarnation claims are made, social status is highly rated and the families involved would have absolutely no reason to make up an inferior past life. Also, in many of the cases there are a variety of witnesses from different families, requiring a conspiracy to exist with no benefit for the individuals.

Fantasy. In this case, the children are making up their stories. Certainly children fantasize a lot about being princesses or superheroes, but a fantasy about a previous normal life would be unusual. In addition, some have accurately described houses they've never seen and people they've never met. In one case, a girl accurately gave the names of 25 people from the previous life. In other cases the children, when first meeting people from their supposed previous family, immediately recognized them and called them by name. Such fantasies would be fantastic in themselves.

Knowledge acquired through normal means. In this case, the children have heard information but have forgotten that they heard it. This is possible in cases involving family members but not with reincarnation claims for strangers. Sometimes these strangers had lived many miles away in a place no family member had been. In addition, why would the children pretend to have been such a stranger?

Faulty memory by informants. In this case, the adults who have reported the cases thought the children gave information that, in fact, they did not. This is a very likely explanation except there are a few cases where the adults wrote or tape-recorded the child's answers, giving a physical record of their claims. In other cases, multiple people have heard the descriptions, and all of them would have to have faulty memories. Also, the children sometimes express an intense emotional longing for their previous family, something that doesn't depend on anyone's memory.

Genetic Memory. Some biologists claim that memories may be able to be transmitted through genes from one generation to the next. In this case, the child would have to be a direct descendent of the person they claim as a previous life, but many of the claims are for unrelated people.

One reason that Stevenson's work has been difficult to refute is that he excluded any claim that had any possibly of a normal explanation. The ones he pursued had to be understood with explanations that weren't normal.

It's also been noted, with skepticism, that so many reports of reincarnation occur in India and Pakistan. One reason may be that both Hinduism and Buddhism include reincarnation in their beliefs, and those religions are strongly represented in both of those countries. Conversely, if an American child told her father, "In my last life my father used to beat me", her parents would probably say she just had a bad dream and to forget all about it. Children normally clam up when they say things that upset their parents, so family discussions about reincarnation would be rare in Western cultures.

Paranormal Explanations

Whenever *para* prefaces a word people get suspicious, like you're trying to hide something or claim something is close to but not quite what you mean. The prefix *para* can have several meanings, including "beside", "nearby", and "beyond". If you were called "a paragon of virtue" you certainly wouldn't complain. If your boss offered you a "golden parachute" you wouldn't turn it down. Rather than becoming paranoid about some words in a paragraph, it's best to understand their intended meaning.

Paranormal means "beyond normal" and, as Tucker says, "some readers may view all such scenarios as absurd." However, those same readers will play the same "lucky" numbers in the lottery or wear the same shirt they wore when they won a previous sporting event to the next event. Bruce Hood, Professor of Experimental Psychology at the University of Bristol, offered students £10 to put on an old blue cardigan sweater and everyone volunteered. But when he told them the sweater had been previously worn by a famous murderer, nearly no one volunteered. The students attributed some evil attached to the sweater, transferring something nonphysical to a piece of clothing. What's paranormal to one person is normal to another.

Certainly, if a situation can't be explained through normal means, we have to look elsewhere. Here are the possibilities:

Extrasensory Perception (ESP). As the name implies, ESP involves detection beyond the physical senses. Three main types of ESP are considered: *telepathy*, *clairvoyance*, and *superpsi*.

With telepathy, one person reads another person's mind, which is more than difficult when the other person died some years before. In addition, the children Stevenson interviewed believe that they really are the previous person, not just that they're reading thoughts about the other person from someone else.

Clairvoyance is exhibited when a person knows facts about another person from holding personal objects from the other person, not by reading minds. However, most of the children didn't have any such objects.

Superpsi means the person is capable of knowing anything that's possible to know. That means that if the knowledge is available anywhere, in any form, even if only in a person's mind, someone with superpsi can know it. So, if someone has some unusual knowledge and no one can explain how they got it, the answer is that they got it through superpsi. The problem with a concept like this is that it can explain anything and can't be disproved, but you already knew that somehow.

We should also keep in mind that we're talking about very young children here, mostly between 2 and 4 years old. These are not young mystics and don't appear to have any paranormal talents in any part of their life.

Possession. In possession, some spirit takes over the body of a living person and that person becomes who the spirit used to be. However, most of these children have only some memories of another life, and those memories start to disappear by the time the child is 6 or 7 years old. As it happens, children almost always lose their early childhood memories when they're 6 or 7, for reasons that are not well understood, in a phenomenon called *early childhood amnesia*. Other children who exhibit preferences and/or phobias characteristic of the previous individual often exhibit them before talking about the previous life, and their parents don't notice dramatic personality changes when the previous life statements begin. The kid seems like the same kid they had before, except now they're talking about having been someone else.

Reincarnation. The last possibility is reincarnation. Jim Tucker points out:

> Our cases contribute to the evidence that consciousness can survive death in at least some situations, and this is surely a more important finding than any specific ones that we may discern. This means that each of us is more than just a physical body. We have a consciousness as well that is capable of surviving the death of that body. If we change the terminology from consciousness to spirit, then we can say that we all have a spiritual component along with our physical bodies. (p. 229)

For our purposes, reincarnation is limited to the elements of the essence, contained in another dimension that's seen by us through our personality and described more fully starting in Chapter 4.

The classic understanding of reincarnation is being reborn as another individual, but I suggest the essence never actually died – it just returned to its dimension when the brain in the physical body died. Consequently, reincarnation for us is really the same immortal essence being hosted in a new physical body.

Why am I spending so much time on reincarnation? Because it offers an answer to the question:

Why is her personality so different?

If we go so far as to accept that reincarnation may occur, other questions immediately come to mind. Are other living things besides humans (dogs, fish, amoeba, pine trees etc.) reincarnated? If not, is there some property in a living thing that allows reincarnation? To address those issues, we have to see where the body contains the essence. We'll start with how life started and has evolved to where we are now.

Chapter 2
Origin and Evolution of Life

This chapter looks at the origin of living things and how life developed over time through evolution. Specific aspects of what makes a living thing alive are discussed.

The Origin of Living Things

There are two different views of the origin of living things. The first is the spiritual view, in which all living things were created by a Supreme Being. The second is the biological view, in which all living things evolve from previous living things. The first view is fundamental to nearly all religions and a typical description of it can be found at the beginning of the Old Testament in Genesis:

> And God said, 'Let the waters bring forth swarms of living creatures, and let birds fly above the earth across the dome of the sky.' So God created the great sea monsters and every living creature that moves, of every kind, with which the waters

swarm, and every winged bird of every kind. And God saw that it was good.

And God said, 'Let the earth bring forth living creatures of every kind: cattle and creeping things and wild animals of the earth of every kind.' And it was so. God made the wild animals of the earth of every kind, and the cattle of every kind, and everything that creeps upon the ground of every kind. And God saw that it was good.

Then God said, 'Let us make humankind in our image, according to our likeness; and let them have dominion over the fish of the sea, and over the birds of the air, and over the cattle, and over all the wild animals of the earth, and over every creeping thing that creeps upon the earth.' So God created humankind in his image, in the image of God he created them; male and female he created them.

The second view, the biological view, removes the Supreme Being from the process and inserts adaptation to the environment as the basis for the creation of new kinds of living things. This view is called *evolution* and is described in the remaining sections of this chapter.

Before we get into evolution, a few points should be noted. The substantial amount of obvious evidence showing that living things physically change to accommodate their environments has convinced spiritualists that evolution does occur, but they draw the line at it occurring for humans.

Also, there are many scientists who are strong believers in God, but the innate conflict between creation by a Supreme Being and creation by biochemical processes is a major issue. Any discussion about it is full of heat from both sides. The spiritual people ask the biological people, "If there's no God, how did life form in the first place?", and the biological people have no answer ("We're working on it" isn't an answer). Similarly, the biological people ask the spiritual people, "Where did God come from?", and the spiritual people have no answer ("God always existed" isn't an answer). The spiritual people are sure the biological people are total idiots who ignore the word of God and are going to hell because of it, and whatever can be done to save them must be done. The biological people are sure the spiritual people are total idiots who ignore everything science has learned about life, and whatever can be done to educate them must be done. Good luck to everyone.

How Life Developed

Life has been on Earth for about 3.5 billion years. Any idea what it's like to have a billion of something? Let's say you lived the average American life expectancy of 79 years and were born with a billion dollars (that's $1,000,000,000). To leave nothing for your ungrateful heirs, you'd have to spend about $35,000.00 every day of your life from birth. Let's try that again. You'd have to spend about thirty-five thousand dollars every day of your life, from the day you were born until the day you die 79 years later, including Sundays and holidays, to spend it all. The ultimate shop-'til-you-drop.

But most people can't appreciate times in billions of years anyway, so let's look at it differently.

Time (Years Ago)	Time Period	Major Events
3,500,000,000	Archaen	Bacteria microfossils appear
2,500,000,000	Proterozoic	Bacteria, algae, and simple animals appear
543,000,000	Paleozoic	Animals increase in diversity
248,000,000	Mesozoic	Flowering plants, birds, and dinosaurs appear
65,000,000	Cenozoic	Today's plants and animals develop
7,000,000		Oldest human
200,000		Modern humans

Figure 2-1. Timeline for Life

As Figure 2-1 shows, humans didn't appear on Earth until about 98% of the "living thing time" had already passed. Paleoanthropologists nicknamed that hominoid "Toumaï", which means "hope of life" in the Dazaga language of Chad in Africa where the fossil remains of the *Sahelanthropus tchadensis* species were found. Modern humans like you and me, of the *Homo sapiens* species, didn't appear until about 200,000 years ago, after more than 99.99% of living thing time. From a biologic time scale (never mind a geologic time scale or a cosmic time scale), we've just been born.

Here's how biological evolution says we got here. Life first appeared about 3.5 billion years ago in the form of bacterial microfossils that, over time, created *stromatolites*. Stromatolites are rocks formed from layers of these bacteria mixed with mud. They're still produced today, and look very much like those found as fossils. All this happened (and still happens) in the ocean.

But where did these bacteria come from in the first place? Theories abound. Some scientists favor the "lightning" theory, in which a bolt of lightning zaps a mixture of inorganic materials and creates something that can reproduce itself. In 1953, Stanley Miller and Harold Urey at the University of Chicago ran an electric current through a flask containing a mixture of gases thought to make up the atmosphere of the early Earth, and created amino acids as a result. Amino acids are the building blocks of proteins and dictate the biological activity of proteins. But they didn't create life.

Your Essence, Your Eternity

Other scientists favor the "hydrothermal vent" theory. Old Faithful geyser in Yellowstone National Park is a hydrothermal vent, but the ones possibly involved in the formation of life occur deep in the oceans. The environment around these vents seems to be totally inappropriate for life. The undersea pressure is enormous, there's no light or oxygen, the water is filled with noxious chemicals from the vent, and the temperature from the vent is around 400° C (the water doesn't boil because of the pressure). In 1976 a group of marine geologists used the research submersible ALVIN to see one of these vents for themselves and found a large variety of life forming a food chain, including bacteria, snails, shrimp, crabs, fish, octopuses, and very strange tube worms. Actually, all the life at a hydrothermal vent is strange. Everything living there depends on the vent for both heat and basic life support and are restricted to the immediate vicinity of the vent for their entire lives. But there it is, life, at a place where no one would expect it.

Biologists classify all life into one of three *domains* – Archaea, Bacteria, and Eukaryota. All life is composed of deoxyribonucleic acid (DNA) arranged as *chromosomes* contained in *cells* with ribonucleic acid (RNA) thrown in. Viruses don't have cells and many are composed only of

RNA, but because they need to infect host cells to generate their metabolism and to reproduce they usually aren't considered to be living things.

Archaea

Originally classified as Bacteria, these single-chromosome organisms can live in extreme environments such as the hydrothermal vents just described, in alkaline or acid waters, in methane, in deep petroleum deposits, in your digestive system, etc. They're also abundant in the plankton of the open sea. Like Bacteria, their DNA isn't enclosed in a nucleus. However, their cell walls contain different kinds of amino acids and sugars than those found in Bacteria, and drugs that affect Bacteria have no effect on Archaea.

Bacteria

Bacteria are single-chromosome organisms with an estimated 5×10^{30} individuals and are easily the most abundant and diverse living things on Earth. They have no nucleus, chloroplasts, or mitochondria, and their ribosomes are different from those in Eukaryota. There are about 10 times as many bacterial cells in your body as there are human cells, mostly on your skin, genital areas, mouth, and intestines. They're the germs that make you sick but they also make vitamins, break down garbage, and maintain our atmosphere. Kill them at your peril.

Eukaryota

For nearly everyone, Eukaryota are the only living things they've ever seen. This domain includes all the plants and animals – mammals, birds, fish, insects, trees, algae, etc. The distinguishing feature is a membrane-bound *nucleus* that contains and protects the genetic material. Eukaryotes were the last to be formed and represent a tiny minority of living things.

Other scientists favor the theory in which organic materials, or even basic life forms, were on meteors or other debris that crashed into the early Earth and survived. In 2011 NASA took a tiny animal called a *tardigrade* (Figure 2-2) into space and exposed it to near-absolute-zero temperatures, boiling water, no atmosphere, and intense radiation. Some of the animals survived and were able to breed later.

Origin and Evolution of Life

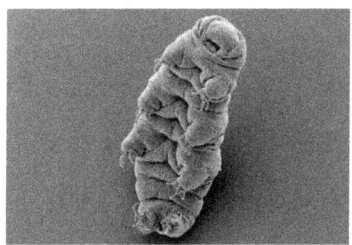

Figure 2-2. Tardigrade
Bob Goldstein and Vicky Madden, UNC Chapel Hill

In the year 2000, a meteor loaded with organic material landed in Tagish Lake in Canada. Because this meteor was so well preserved, NASA scientists were able to slice it and scan it with an electron microscope. They found submicroscopic globules containing mostly carbon, oxygen, nitrogen, and hydrogen, formed in a way that requires nearly absolute zero far away from the Earth. While the globules aren't alive, they are organic and contain the necessary ingredients for life. Of course, if actual living things were on any meteors, it naturally begs the question of where that life came from.

Building Blocks of Life

Organic compounds are necessary components of life, but they aren't alive themselves. There are chemicals and compounds of all sorts in living things – what causes them to be made, and why? How do living things reproduce to form new living things? There must be a conductor telling this huge and complex orchestra how to play – what is it?

There are two prime suspects – RNA and DNA. Both seemed to be involved whenever a cell is replicated into a second cell. The main structural difference between the two is that RNA is usually a single strand of components while DNA is usually a double strand. In the 1950s James Watson, Francis Crick, and their colleagues figured out that replication was done by unwinding the double helix of DNA and copying it into a second double strand, which then separated into its own cell. RNA helped as a messenger to trigger reactions and bring necessary components to the DNA.

Figure 2-3 shows the general structure of the DNA double helix. The backbone is made of a sugar phosphate with each strand connected by a pair of chemical bases. There are four bases in DNA – adenine, thymine, cytosine, and guanine. Because of their chemical structure, the strongest bonds are when adenine connects with thymine and cytosine connects with guanine. When DNA is replicated each adenine finds another thymine, and so forth. Biochemists always refer to these bases by their first letters

(A, T, C, G), so if you hear someone using those letters together s/he is probably a biochemist. Smile and nod knowledgably.

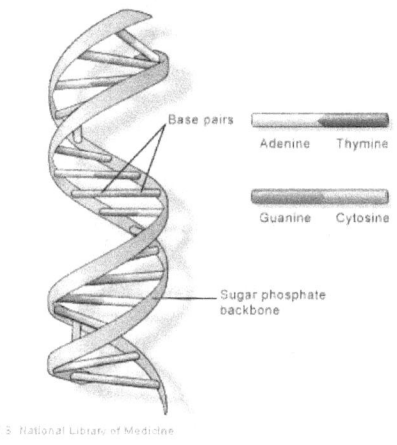

Figure 2-3. DNA Structure
From the U.S. National Library of Medicine

When life was first developed RNA was probably the main replicator because it's relatively simple. Unfortunately, it's also relatively unstable, so it was overcome by the more stable DNA that developed proteins to direct cellular functions, with RNA as a helping messenger. How does DNA do this?

By your *genes*, which are groupings of the base pairs on the DNA strands. You can think of genes like words constructed of individual letters (the base pairs). Genes specify your eye color, how tall you'll grow, if and when you'll go bald, and generate all the chemicals that control your bodily functions. Humans seem to have about 20,000 genes to do all these things. You might be surprised by this rather small number, considering that rice and corn also have about 20,000 genes. Oh, well.

Beginnings of Life

Irrespective of the way in which life originally formed, the process that creates the variety we see is biological evolution. At first, simple compounds were able to replicate themselves, sometimes with errors. As we see today, replication errors are almost always bad and are usually fatal, and this was probably the case with the early replicating molecules. But occasionally a replication error permitted the compound to improve its situation, perhaps able to withstand a harsher environment or use alternate energy sources, and that new compound pushed aside its predecessor. This process is called *natural selection*. And so it went. At some point, a replicating molecule enclosed itself within a membrane, which gave it two huge advantages. First, the replication process could be better protected and second, the enclosed part could be different from the external environment. This membrane became what we know as a *cell membrane*, and was such an improvement over the naked compounds that it quickly outselected them. Eventually the compounds became complex enough to be recognizable as life forms like bacteria, with parts of them performing different metabolic functions. After awhile, cells became differentiated with some cells better equipped to perform certain functions, giving rise to multicellular living things. An example of a multicellular organism is red algae, found as fossils that are 1.2 billion years old.

All these early living things relied on diminishing amounts of existing organic molecules. Hunger is a terrific motivator, and some cells adopted sunlight as an energy source. This had an enormous impact on the development of life on Earth. By using photosynthesis to produce energy the plentiful amount of carbon dioxide and water were the raw materials, while oxygen was the waste product. Although each photosynthetic cell produces only a minute amount of oxygen, a large number of cells over a long time changed the Earth's atmosphere to the one we enjoy today. But the living things at the time didn't enjoy oxygen, which was toxic to them. They, too, either died from oxygen poisoning or evolved to change their energy systems to use this new material. And some of this oxygen was used to form the atmospheric ozone that protects the Earth from the sun's ultraviolet radiation.

The sun's ultraviolet radiation made life on land very difficult, so early life was restricted to the ocean. The ultraviolet protection provided by the ozone layer meant that organisms that reached land were less likely to die. The first land plants to thrive were probably algae and fungi, something like 500 million years ago. Plants and animals continued to evolve to different types in both the oceans and on land. Insects developed about 450 million years ago to take advantage of this new food source on land. About 375 million years ago amphibians developed to enjoy the new plants and the new insects. To give themselves a wider range and better chance for survival, plants developed seeds. To give their embryos a longer gestation period and consequent better chance for survival, land animals developed the egg. Expansion into other parts of the world led to the development of birds and reptiles. Mammals developed about 200 million years ago, although they were all small creatures similar to shrews. A major advantage of

Origin and Evolution of Life

mammals, of course, is their warm blood, which permits them to react more quickly and efficiently in colder conditions. Flowering plants first came about 132 million years ago. By 32 million years ago some land mammals had had enough and returned to the sea, to become whales and dolphins. Humans didn't separate from our closest relative, the chimpanzee, until about 3.6 million years ago.

While the development of life seems to be active and continuous, keep in mind that we're talking about hundreds of millions of years and a lot can (and did) happen along the way. Supercontinents were formed and broke up, volcanism was very active, ice ages came and went, and the Earth was occasionally hit by substantial meteors. Major extinction events included one about 488 million years ago, of unknown cause, that wiped out 49% of the species on Earth; one about 364 million years ago, possibly caused by an asteroid hitting in either Nevada or Australia, that wiped out at least 75% of species on Earth; one about 250 million years ago, possibly caused by an asteroid hitting in either Australia or Antarctica, wiped out 95% of the marine species and 75% of the land species; one about 200 million years ago, of unclear cause, wiped out about 50% of all species; and one about 65 million years ago, possibly caused by an asteroid or comet impact off the Yucatan Peninsula, wiped out about 30% of all species, including the dinosaurs, and was the start of the rise of the mammals. All of these events included atmospheric change in which debris kicked up by the impact and/or associated volcanism blocked the sun's rays, resulting in photosynthetic plant die-off and consequent food chain disruption and years of very cold temperatures.

Cold temperatures, called ice ages, also came from other sources. An event about 770 million years ago was so severe that the surface of all the oceans completely froze, and was only broken by volcanism that spewed enough carbon dioxide into the atmosphere to cause a greenhouse effect. An ice age was part of the problem in the major extinction 488 million years ago. Some scientists say we're in an ice age now, using the argument that through most of history Earth has been ice-free, even at high latitudes. One accepted definition of "ice age" is when ice sheets exist in the northern and southern hemispheres. Since we still have ice sheets in Antarctica and Greenland, we could be in an ice age. If so, it began about 40 million years ago and included a glacial period that ended about 10,000 years ago. That was when the Mammoths and Saber Tooth Cats disappeared and is well within the human experience.

Living Things

The world is full of living things. To understand how life works it's helpful to organize the various living things according to their characteristics. The Swedish botanist and medical doctor Carl Linnaeus looked at the similarities among plants and animals and published *Systema Naturae* in 1735, in which he outlined his method of classifying all living things. His categories of *genus* and *species* are still used today. But until Charles Darwin published his theory of evolution in 1859, naturalists spent their time identifying and naming newly discovered plants and animals without any thought to how their similarities and differences might group them. This was due to the belief that all living things had been created by God as they were, so there was no need to look for any ancestors or relationships among them.

Figure 2-4 shows the hierarchy used to classify living things, with the more general groupings towards the bottom and the more specific groupings towards the top.

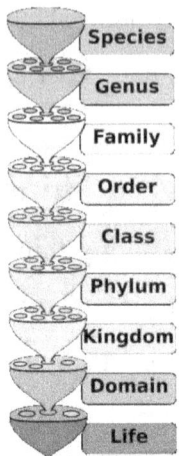

Figure 2-4. General Classification of Living Things
Drawn by Peter Halasz

The General Classification was created as a formal structure by biologists with all the names in Latin. For example, the genus containing dogs is *Canis* and the genus containing humans is *Homo*. The most common specific species have English names in addition to their scientific Latin names. Dogs are *Canis familiaris* and humans are *Homo sapiens*.

Considering the scope and complexity of living things, it's not surprising that various distinguished biologists have proposed different methods to classify life. Some classifications are based on their physical characteristics (size, shape, etc.), some on their functions (breathing, movement, etc.), some on their DNA or RNA structures, etc.

It's estimated there are 8-13 million different species on Earth, of which 5-10 million are bacteria. There's only one human species. The completeness of the count depends on the interest in the research community, and is more accurate in some groupings (such as mammals) and likely very inaccurate for others (such as insects) but, as biologists point out, more than 99 percent of all species ever living have become extinct.

You might think we know what lives on Earth, but from time-to-time new species and even new forms of life are discovered. Usually these are types commonly known as *extremophiles* because they thrive in environments too extreme for other living things. Examples are the tube worms and crustaceans around the undersea volcanic vents who tolerate very high temperatures and water loaded with sulfur; microbes that live two miles deep in a freshwater lake in Antarctica; glass-eating microbes living in volcanic rock as far down as 4 miles (2.5 km) below the ocean floor; and microbes in the Earth's crust that eat hydrogen and exude methane. These environments are similar to those found on other planets, expanding the possibility that life could exist in places that are too harsh for most living things.

But you don't have to go to extreme environments with a microscope to find a new species. In 2010 scientists discovered dozens of new species, including 50 new types of spider, a striped gecko, and three frogs in the forests of Papua New Guinea. The biology of Earth is complex and evolution continues – who knows what you'll find in that next walk in the woods (or under your sink)?

When a person who isn't trained in biological classification sees a living thing, it's hard to classify it accurately. If you see a furry little animal scurrying around, is it a mouse, a gerbil, a hamster, or something else? They all look rather alike. But not only are they different species, they're also in different genera (the common house mouse is *Mus musculus,* the gerbil is *Meriones unguiculatus*, and the hamster is *Cricetus cricetus*, with the genus as the first word and the species as the second word). If we're to understand living things, we have to understand their differences.

Ever wonder why we can't cross a potato with a pineapple to get potatoes that grow on bushes so we don't have to do all that digging? It's because they're different species. Only individuals within the same species can reproduce naturally. An exception is when scientists take DNA from one living thing and put it in another to produce living things that don't naturally occur, but normal reproduction requires individuals in the same species.

We should also be clear about what being a "species" means. There are about 20,000 species of butterflies. There are seven species of canines (dogs, wolves, foxes, coyotes, dingoes, jackals, and African wild dogs). *Breeds* and *varieties* are not species. So, for example, Irish Wolfhounds and Chihuahuas are both dogs and can interbreed (although they might have a hard time), but an Irish Wolfhound and a Gray Wolf cannot. Remember, there's only one human species, so all the varieties we see in humans are just that, varieties. We're all the same single species.

Overview of Evolution

Evolution means change over time. If you visualize clouds forming or mountains eroding, you're visualizing evolution. But that kind of evolution isn't the same as biological evolution. Biological evolution is the change in living things, through changes in their genes, over time. It's biological evolution that's of interest to us.

Before Charles Darwin published *On the Origin of Species* in 1859, all living things were considered static creations of God. There were certainly variations in living things, such as breeds of horses or colors of roses, but nothing ever changed, on its own, to something else. Anyone who said anything else was a heretic. In those days, if there was one thing you didn't want to be called, it was a heretic.

Actually, Darwin developed the basis for his work in 1838 but was concerned that such radical ideas required substantial additional material to back them up, to avoid being labeled a radical himself. He continued his research but, in 1858, he received a paper from another biologist, Alfred Russel Wallace, which included preliminary ideas similar to his own. To be sure he would receive credit for his theory, he published and held his breath.

Reaction was swift and predictable. The public loved his ideas and the church and scientific community hated them. Darwin considered the criticism of eminent British physicist William Thompson (known as Lord Kelvin) as especially damaging. Thompson said that the natural selection of evolution, which required billions of years of incremental changes to explain the diversity of life on Earth, was impossible because the sun could be no more than a few tens of millions of years old. Thompson thought the sun was made of molten rock, and it would be another 70 years before spectrographic studies would show it's really mostly hydrogen.

Darwin's entire stock of books was oversubscribed when they went on sale, even though he avoided any inference to evolution related to humans (his later book, *The Descent of Man*, expanded his ideas to the ones we know today). The position of the church was (and is) somewhat parallel to the ecclesiastical position presented against Galileo when he suggested that the Earth was not the center of the universe. Fortunately, unlike Galileo, Darwin didn't have to spend the rest of his life under house arrest.

As it happens, it fell to an Austrian monk, Gregor Mendel, to correctly describe inheritance. At the same time that *On the Origin of Species* was published, Mendel was experimenting with peas to understand species variety. Often called "the father of modern genetics", Mendel realized that genes were the reason for genetic variability, although others (including Darwin) believed that pangenes were responsible. Basically, the pangene theory is that every body cell has a vote in determining the characteristics of offspring. But we're concerned more with evolution than inheritance, so perhaps a short description of how biological evolution works would be useful.

Biological Evolution

Biological evolution is the process by which new living things are descended from ancestor living things, but the new living things are different. These changes are permanent because they're changes in the genes. The best way to illustrate biological evolution is through examples, and a classic example is the peppered moth.

Peppered moths try to hide for much of their life from the birds that find them so tasty. Since the moths live around trees, those that blend in with the trees are better hidden and have a better chance for survival. Those with colors different from the trees are more easily spotted by birds and are eaten.

Historically, nearly all peppered moths had light coloration, which effectively camouflaged them against the light-colored trees on which they rested. Figure 2-5 shows the coloration of a typical peppered moth.

Figure 2-5. Lighter Colored Peppered Moth
(courtesy Olaf Leillinger)

Not all peppered moths have such good coloration. Some are darker, such as that shown in Figure 2-6.

Figure 2-6. Darker Colored Peppered Moth
(courtesy Olaf Leillinger)

As you can imagine, those darker moths were found much more easily by the birds and, consequently, fewer of them lived to reproduce. However, pollution during the industrial revolution in England caused many of the trees to be covered with dark soot, reversing the situation. Then, the dark moths were more effectively hidden and the lighter ones became bird food. This caused a bloom in the numbers of darker moths and a large reduction in the numbers of lighter moths.

In recent decades, however, due to improved environmental standards, the trees are light again and the moths have become lighter again. The moths practice *differential reproduction* based on variations in their habitat, resulting in a new heredity for their offspring. This example of *natural selection* is one of the foundations of biological evolution.

Another way biological evolution can occur is through *mutations*. A mutation is a change in an organism that creates a totally new type of organism. Such changes happen when cells divide but a mistake (or more) occurs in the DNA replication, producing a different result in the new cell. Mutations are a common and natural occurrence in cell division, but other factors (such as drugs or radiation) can foster them. Figure 2-7 shows a mutated pig with two snouts, two mouths, and three eyes in its head. Most cell mutations are harmful, causing the cell to die; if such a mutation occurs in a germ cell (those used for reproduction), the offspring will die. However, some mutations are beneficial to the living thing, and produce an improved offspring.

Figure 2-7. Mutated Pig
(photo by Reuters)

Why don't we see these improved offspring? We would if we looked in the right places. Mostly we look at large creatures with relatively long reproductive cycles, but effective mutations usually need a lot of time. Nature has had that time (over 3 billion years) to fiddle around, but our short lifetimes keep us from seeing these changes. Medical researchers see mutations all the time, unfortunately, in diseases that become resistant to previously-effective drugs. Bacteria and viruses reproduce quickly and consequently the duplication mistakes made in their DNA occur much more frequently. So, if a genetic mistake happens to make a bacteria or virus resistant to a drug, that bacteria or virus has a much better chance of reproducing. Then researchers must find a new drug it's not resistant to. It's the old cops-and-robbers game, with the robbers able to keep one step ahead.

Your Essence, Your Eternity

But biological evolution says there have been improved offspring, and we're one of them. As mentioned, before about 3.6 million years ago no living things that could be considered "human" existed. Then a mutation occurred in one of them that gave it the ability to walk upright, at least for awhile. This was a distinct advantage, permitting more extensive use of the arms and hands for manipulating objects, food collection, defense, and other purposes. Lucy, of the species *Australopithicus afarensis*, was one of them, and is classified as the first stage towards human. True upright stature wasn't attained until *Homo erectus* came around, about 1.5 million years ago. *Homo erectus* had mutations that included locking knees and a different location of where the spine attaches to the brain, permitting the individuals to stand upright continuously.

Neanderthals (*Homo neanderthalensis*) developed about 350,000 years ago. Their mutations seem especially selected for the cold weather of the time, with robust bodies and large brains. Actually, their brain size was larger than ours, they were about our size, and they were much stronger. Some researchers consider them a subspecies of our own, *Homo sapiens*. While *Homo sapiens* came later, Neanderthals were still around but there's no firm evidence they could breed together. Consequently, they're considered separate species by most researchers. There's evidence of Neanderthals burying their dead, sometimes with tools and implements, which strongly implies that they had funeral ceremonies and associated spiritual thoughts. The last Neanderthals known lived about 24,000 years ago.

Which brings us to us, *Homo sapiens* (in case you're wondering, *sapiens* means wise or intelligent). Our species started about 200,000 years ago with mutations including reduction of the canine teeth, modifications to the larynx (making speech possible) and slower development of postnatal brain growth (allowing an extended period for social learning). The associated evolution of complex social structures may be related to the brain reorganization found in *Homo sapiens*. As might be suspected, there's plenty of controversy about the cause and effect of many of our attributes, but we're the only species we know that has overt spiritual interests.

Evolutionary psychologists suggest that all human behavior, including language, creativity, and morality, have developed through natural selection processes. The underlying concerns are if living things exist solely because of the natural selection described by evolution, how are humans special and what does morality mean? It's tough to accept a world in which biology determines everything.

The kinds of judgements we make reflect our personality and our essence, but those judgements are based on the information our senses receive. Before we can understand our essence, let's understand a bit more about how we sense things.

Chapter 3
Sensing Our Habitat

This chapter helps us understand what we know, or think we know, about our habitat. We discuss reality, how our five senses collect information about reality, and how our reality is different from that of other living things.

Reality

What is reality?[1] When you think about it, the question really is, "How do I interpret reality?" Our physical body includes various receivers to bring information from the outside world to our attention. Those receivers are our five senses of sight, hearing, smell, taste, and touch. All you know about anything is the result of inputs from those senses that are processed by your brain. If an image is too bright, a sound too loud, an odor tantalizing, a taste delicious, or if something feels rough, those conclusions were determined by your brain, not by your senses. To

---

[1] Most of the data referring to living things in this section was obtained from public educational Web sites maintained by the University of Washington and by the Oracle Education Foundation.

further impress the point, your eyes are just fancy lenses that don't see a thing; your ears are just fancy microphones that don't hear a thing; your nose is just a fancy gas chromatograph that doesn't smell a thing; your taste buds are just fancy chemical receptors that don't taste a thing; and your nerve endings are just fancy stimulus receptors that don't feel a thing. All recognition of stimuli, no matter what they are, is done by your brain. And, if you and your identical twin are standing next to each other, each of your interpretations of the world may be different because your senses may have received their inputs differently. Reality is a very personal matter.

Your Five Senses

Your reality is also a very restricted matter. Consider your sense of sight, which begins by your eyes collecting photons from the world outside of your body. Not all photons are collected – only those with wavelengths between 380 and 760 nanometers (billionths of a meter). Figure 3-1 illustrates how little of the total electromagnetic spectrum our eyes can see, with resulting restrictions of our view of the universe.

Sensing Our Habitat

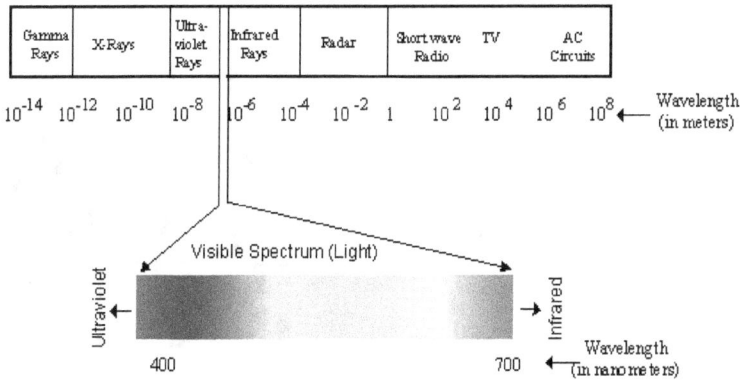

Figure 3-1. The Visible Spectrum

Speaking of the universe, Figures 3-2 and 3-3 show one effect of our limited vision. Both images were taken of the Carina Nebula, an area of dust and gas with new stars forming inside, by the Hubble Space Telescope. Figure 3-2 shows the nebula in visible light as mostly obscured by the dust. Figure 3-3 was taken in near-infrared light, which is beyond our visual ability but which sees through most of the dust and clearly shows two narrow jets of energy blasting from a still-hidden infant star.

Figure 3-2. Carina Nebula Visible
Courtesy of NASA

Figure 3-3. Carina Nebula Infrared
Courtesy of NASA

The Space Telescope is really a big eye-in-the-sky, with a diameter of 2.4 m (7 ft 10 in) compared with your eye's diameter of about 0.024 m (0.94 in). A bigger lens will certainly let in more light, but the real key to sharper vision is the number of sensors ("cones" in the eyes of living things) the light stimulates. Humans have about 4.5 million cones. The hawk, with a smaller eye, has many more cones and consequently has vision that's about eight times sharper than a human's vision. How far away does a small rodent have to be before your girlfriend screams? Buzzards can see small rodents from a height of 15,000 ft (that's almost 3 miles). Some insects have complex eye structures that permit specialized vision and may allow them to detect electric fields associated with light (for example, ants can see polarized light), and some fish can see in the infrared or in the dim bioluminescence of the deep ocean. Penguins can see in the ultraviolet. Our limited human vision misses a lot of what's out there.

Our vision misses even most of what we see. Marcus Raichle points out that about 10 billion bits per second arrive on our retina, but because our optic nerve only has a million output connections just six million bits per second leave the retina. Only about 10,000 bits per second actually make it to the brain's visual area, and further processing reduces that to less than 100 bits per second. Raichle suggests that some kind of intrinsic brain activity that he calls the brain's *dark energy* must be involved for us to have any kind of perception at all.

How about your hearing? Once again, limitations are set by the hearing apparatus in our ears, which is stimulated by tiny hairs connected to nerve cells. The lowest tone humans can normally hear is about 20 cycles per second and the highest tone is approximately 20,000 cycles per second. You hear best for sounds between 1,000 and 4,000 cycles per second, and your hearing deteriorates with age.

Some researchers say dogs can hear up to 100,000 cycles per second, which is why humans are deaf to dog whistles but they bring the pooch on the run. Bats and marine mammals can emit very high frequency (about 120,000 to 150,000 cycles per second) sounds that are used for echolocation of objects. Susan McGrath reports that some plants have evolved to reflect sound to make it easier for bats to find them. Marine mammals and some land animals (such as elephants) use very low frequency sounds (as low as 0.25 cycles per second) for communication. Crickets have a sound detection membrane in their front legs. And snakes, which have no external ears (sound probably travels through the bones in their head to their middle ear), probably respond to the movement of snake charmers rather than to the sound of the flute (Figure 3-4).

The fakir's banjo failed to have a hypnotic effect on the cobra.

Figure 3-4. Charming a Snake

Sensing Our Habitat

Your sense of smell can distinguish over 10,000 different odors from molecules detected by your nose, but compared with other living things our sense of smell stinks. For example, while we have more than 5 million olfactory receptors in our nose, a dog has more than 220 million, making their sense of smell a thousand times better than ours. Be grateful – would you want the job of sniffing bags for contraband at airports?

Our limited olfactory sensitivity is probably a byproduct of the development of our larger brains – animals in the wild need a sensitive sense of smell for many things including finding food, avoiding predators, and choosing a mate. Mosquitoes are attracted by human body odor (especially foot odor), carbon dioxide, body heat, and body humidity.

Your sense of taste works like your sense of smell – by classifying molecules detected by the taste receptors into fixed categories. The taste receptors can detect five primary sensations – salty, sour, sweet, bitter, and umami (a taste associated with protein foods like meat and cheese). You have between 2,000 and 8,000 receptors, called "taste buds", located on your tongue. Each taste bud holds 50-100 cells that represent all five primary sensations. For comparison, pigs have about 15,000 taste buds, rabbits have about 17,000, catfish have about 100,000, and the entire body of earthworms is covered with taste buds. Some sharks can detect fish extracts at concentrations lower than one part in 10 billion.

The process of tasting isn't restricted to your tongue, or even to your mouth. Michael Moyer points out that taste receptors have been found in the nose and small intestine. Visual and auditory clues also play a part in how something tastes. For example, potato chips taste crisper if you hear a crunch over headphones and white wine with a drop of red food coloring tastes like red wine, even to experienced wine tasters.

While your senses of sight, hearing, smell, and taste are located in specific parts of your body, your sense of touch is done by nerve endings throughout your skin, just under the top layer, in a layer called the *dermis*. The nerve endings send messages to your brain where the feeling is registered. The major feelings registered are those for heat, cold, pain, and pressure. Some areas of your body, such as your tongue and your fingertips, have more nerve endings and are consequently more sensitive to changes in touch. That's why biting your tongue hurts so much and why blind people can read messages written in Braille so quickly just by running their fingertips over the text. Other very sensitive body areas are your hands, lips, face, neck, and feet. The least sensitive part of your body is the middle of your back.

Other animals use their sense of touch in ways useful to them. Butterflies have hairs on their wings that detect changes in air pressure; grasshoppers have hairs to detect air movement; rattlesnakes use their sense of touch to feel the body heat of other animals; and many animals have whiskers to give their sense of touch more distance and more sensitivity. The weird-looking star-nosed mole has 100,000 nerve fibers in the star, almost six times the number in a human hand (Figure 3-5). And, you'll be unhappy to hear, cockroaches are so hard to catch because they can detect movement as small as 2,000 times the diameter of a hydrogen atom.

Figure 3-5. Starnosed Mole

Other living things also have other senses. Turtles and some birds navigate using the Earth's magnetic field. Bacteria and some plants communicate using chemical receptors. Reality is different for different living things.

Personalizing Reality

Your reality is based on the input from your senses and is only your interpretation of what reality actually is. Sometimes your interpretation is wrong. Since your five senses only bring information from the outside world to your brain for analysis, conditions in your brain may deliver an interpretation that isn't real. This happens for people stricken with schizophrenia, but extreme physical body, culture, or habitat conditions may cause your brain to misinterpret the information it gets. You may have also experienced misinterpretations purposely caused by magicians or people doing slight-of-hand tricks. You can't believe everything you see, hear, smell, taste, and/or touch.

Another person's interpretation of the same reality might be similar to or much different from yours. Humans, being the egomaniacs we are, try to impose our views of the world onto others but usually recognize when we're doing this because the people we're trying to convince talk back to us with words not suitable for scientific analysis. That same egomania has us impose our reality onto other living things that don't talk. When you think about it, anthropomorphizing reality makes no sense because the sensory information is different and the subsequent brain interpretations will necessarily be different.

Consider the cicada, specifically the North American genus *Magicicada* that has a 13- or 17-year life cycle. This bug spends nearly its entire life underground, where it burrows looking for root juice to eat. At the end of their lives they burrow to the surface where the nymph sheds its skin and emerges as a flying adult where it and millions of others make a rather loud mating sound that entomologists have the nerve to call a "song". During this time the cicada seems to be oblivious to everything except making noise and mating. You can flick them off a surface and they will simply resume what they were doing. The universe is almost invisible to them.

So, in summary, reality for a cicada is to burrow to find root juice and then finally dig to the surface, shed its skin, make an annoying noise, mate, and lay its eggs. That's it. They know nothing about anything else going on in the world, never mind the math, science, philosophy, and history that interest you so much. Sherlock Holmes notes in *A Study in Scarlet* that he does not know the Earth revolves around the sun, as such information is irrelevant to his work. Cicada reality is what the bug needs to know, it's as right for cicadas as your reality is for you, and neither is better than the other. Living things are all different and their realities are all different. Live with that reality.

Essence Location

Until relatively recently, each human essence was a component influenced by the physical location of the individual. Like-minded people tended to live together, forming those concepts through their leaders and, when necessary, from those in control. Neighborhoods, cities, sometimes whole countries were oriented in parallel ways. The development of inexpensive long-distance travel changed that, permitting individuals with a wide variety of ideas to come and go, often creating dissonance as they went. The availability of fast world-wide communications, especially social media, has also had a significant effect, allowing people to see and experience alternative ideas. The jury is still out on whether all this has been good or bad. It has, however, permitted individuals everywhere to see alternatives.

Our main interest is the interaction between the physical body and essence. The essence is in another dimension and is linked to by the brain to the extent that the brain can handle its capabilities. The brains of other living things link a smaller portion of the essence, with humans having the largest essence. But there's still room to grow, and as humans evolve perhaps their brains will be able to link more aspects of the essence. Those aspects may be similar to the ones described for aliens with big brains in science fiction stories. That could be your physical body in the future. You'll have to wait and see.

We've talked a lot about the essence being in the brain, but we still don't know specifically where in the brain it is. Let's find out.

Your Essence, Your Eternity

Chapter 4
Essence and the Brain

The essence is not part of our physical body but is linked from a dimension outside of our 3-dimensional space and is associated with the brain. In addition to its control center features, the brain functions as the linking area for our essence because the characteristics of the essence require a complex processing ability found only in the brain. Living things that have essences also have attitudes, emotions, morals, and personalities. Only living things that have such characteristics have an essence – living things without them function only with consciousness and biochemistry.

But perhaps it's wrong that the essence requires a brain as its container. Perhaps it's in everything living, albeit with less complexity in less complex living things. Since one of the initial conjectures is that each living thing with an essence has a unique essence, if every living thing had an essence there would be trillions of essences, most of which would be associated with bacteria. If every one of them linked to an essence the reincarnation of a human essence to another human would be impossibly remote.

Brain

Discussing brains is controversial because it's not easy to say if a living thing has a brain. All animals have nervous systems (some researchers say plants do, too), but a brain is generally considered to be a collection of nervous system cells that direct the functions of the whole body. How many cells is enough for such a collection? Specifically, since living things evolve more and more complex nervous systems, at what point is the collection of ganglia considered a brain? While we're most familiar with the human brain, we should recall that brains come in all shapes and sizes and most animals have them.

Another controversial topic when discussing brains is the place of the more ethereal aspects associated with brains, such as morals, personalities, and emotions. Never mind the question of whether or not other animals have such capabilities in their brains. Where do they reside in human brains?

Essence and the Brain

There are two main schools of thought – *monism* and *dualism*. Monism basically means that everything is together in one place. In the case of brains, monists believe that everything is done through biochemistry. True, they say, we don't understand how personalities are developed from the various proteins, hormones, and electrical connections within the brain, but we'll eventually figure it out.

Nonsense, say the dualists. In dualism, the physical component of the brain is separate from the mental qualities. True, they say, we don't know how any of the mental qualities develop or where they are, but we'll eventually figure it out.

Which do you believe? Both philosophies originated with distinguished Greek scholars and both have had equally distinguished champions throughout the ages. However, as is evident from the components of living things graphically shown in Figure 1-1, the separation of the physical body (including the physical brain) from the essence (including personality) is a cornerstone of explaining how ethereal experiences happen. Those experiences, and personality, seem beyond neurochemistry. Some people say if you really believe that intense love is only neuronal activity and an increase in hormones you've never been truly in love. As we look at these experiences in more depth, we'll see more clearly why a dualist view is the most consistent with observations.

Since the essence is associated with the brain, only living things with brains are capable of expressing the qualities of the essence. To better understand how this works, we'll look at the physical structure of the brain and see how it interacts with the essence.

Brain Structure

We've mentioned the difficulty in saying when a living thing has a brain and when it doesn't. Here we'll compare the structures of various brains with each other. This should help us understand what it takes for a brain to be the link for the essence.

Let's start with the nervous system of the nematode *Caenorhabditis elegans* (Figure 4-1). This little worm is one of the most studied creatures on Earth, with Nobel Prizes in Physiology or Medicine for the genetics of organ development and programmed cell death awarded in 2002 and for the discovery of RNA interference in 2006, and the Nobel Prize in Chemistry for work on green fluorescent protein awarded in 2008, all for work with *C. elegans*. In addition, specimens aboard the Space Shuttle Columbia survived the disintegration of that spacecraft in 2003.

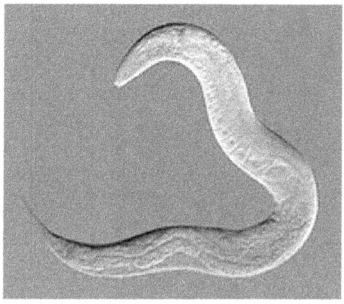

Figure 4-1. Nematode *C. elegans*

The nervous system of *C. elegans* contains a total of 302 neurons and the connections between all of them have been mapped (Figure 4-2). However, as Carl Zimmer notes, "investigators still do not know how that simple network gives rise to a working nervous system" (p. 63). Obviously there's no room for a personality and no link to an essence.

Your Essence, Your Eternity

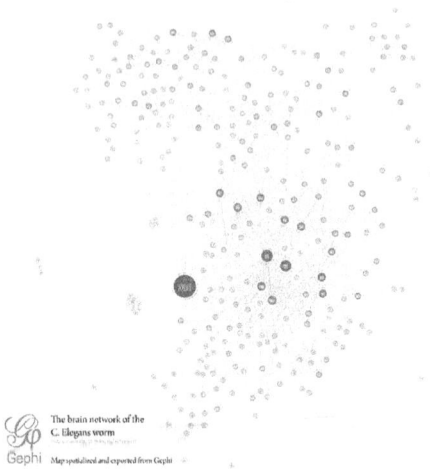

Figure 4-2. *C. elegans* Neuron Connections
Data computed by D. Watts and S. Strogatz; Map spatialized from Gephi

But nervous system complexity goes far beyond its structure. Figure 4-3 shows a computer simulation of the communication between two neurons across a *synapse* – the point of contact between two nerve cells. The communication is done when one cell ejects a subset of over 1,400 types of molecules as neurotransmitters (dots in the figure between nerve cells). Carl Schoonover describes how such simulations of moment-by-moment and molecule-by-molecule may help us understand how messages are processed in the brain. We're just getting started in even appreciating the intricacies of neuronal communication, so your own brain synapses will have to be patient.

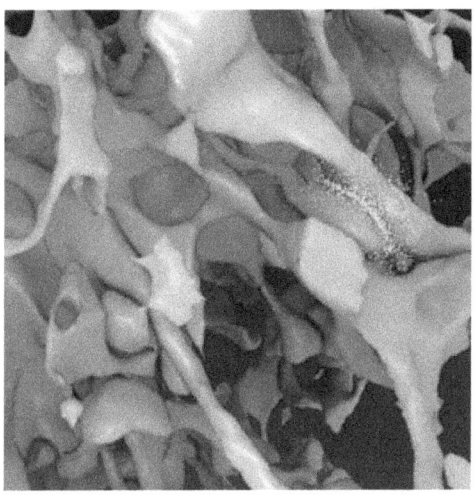

Figure 4-3. Neuronal Communication
Image generated by Tom Bartol Salk Institute for Biological Studies in collaboration with Justin Kinney, Dan Keller, Chandra Bajaj, Mary Kennedy, Joel Stiles, Kristen Harris and Terry Sejnowski

Figure 4-4 shows the brains of various mammals. To give some sense of mass, Figure 4-5 shows the approximate weights of each of them. Don't feel too pompous – some animals have larger brains than humans. As it happens, the brain weight of the average adult human male is 1400 grams (3 lbs) while that of the average adult human female is 1300 grams (2.9 lbs). I don't know about you, but I'm not going to tell my wife I'm smarter than her because my brain weighs more. Also, while humans have a very large brain weight to body weight ratio (2.26%), the hummingbird's is 4.2%! But don't give up – neuroscientists are hard at work trying to find a ratio that proves we're the smartest species on Earth.

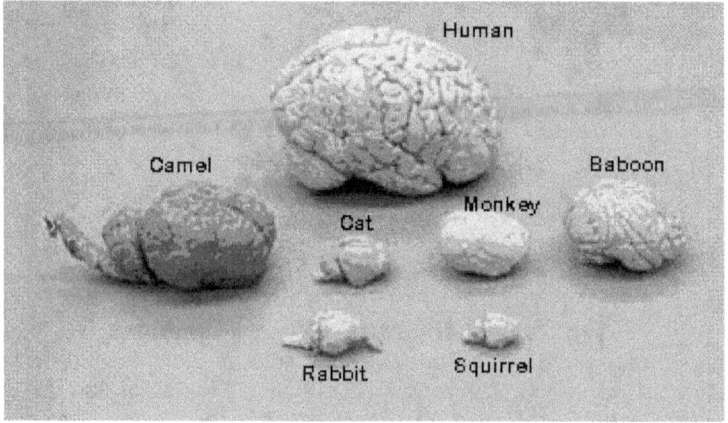

Figure 4-4. Brains of Various Mammals
From Bryn Mawr College

Most anthropologists tend to agree with Darwin's proposition that "the difference in mind between man and the higher animals, great as it is, certainly is one of degree and not of kind."

Brain	Approximate Brain Weight	Approximate Body Weight	Brain Wt / Body Wt
Human	1400 grams	62,000 grams	2.26%
Camel	680 grams	529,000 grams	0.13%
Baboon	140 grams	30,000 grams	0.47%
Monkey	100 grams	7,000 grams	1.43%
Cat	30 grams	3,300 grams	0.91%
Rabbit	12 grams	2,500 grams	0.48%
Squirrel	6 grams	900 grams	0.67%

Figure 4-5. Brain Comparisons
From Bryn Mawr College

There are some immediate physical similarities and differences you can see in the various brains. All of them have two hemispheres and there seem to be distinct structures in each brain. Also, each brain has a spinal cord leading from it. Besides the size difference, some of the brains have a lot of complex folding while others appear rather smooth. The extent of folding doesn't seem to directly correspond with intelligence or behavioral complexity. For example, the brain of a dolphin or whale is more convoluted than that of a human. So what does it mean? Hard to say. Different neuroscientists have found different relationships, but nothing definitive has been accepted by everyone. While we're still in the infancy of brain research, Yuste & Church note that focused technology development has helped to make exciting discoveries happen. They say,

> Such breakthrough methods could, in principle, begin to bridge the gap between the firing of neurons and cognition: perception, emotion, decision making and, ultimately, consciousness itself. (p. 40)

Most research, naturally, has focused on the human brain. "We have not really been able to understand why the human brain is so much more capable than that of any other animal," said neuroscientist Maiken Nedergaard. "Some people have thought that it's simply that a bigger brain is a better brain, but an elephant's brain is bigger than a person's, for example, but it's not nearly as powerful." Some researchers think the complexity of the human brain is the key, but we have to find that complexity first.

Let's take a look at some of the most interesting parts of our brain.

Human Brain Structure

Figure 4-6 identifies major sections of your brain. The outside ¼ inch covering both hemispheres is the *cerebral cortex* (sometimes called the *gray matter*) that controls your thinking, voluntary movements, language, reasoning, and perception. "Cortex" is the Latin word for "bark" to indicate the outside covering. Cerebral cortex sounds so much better than cerebral bark, don't you think?

The cerebral cortex is also the general term for all the "folded" parts of the brain and is divided into four sections, called *lobes*: the *frontal lobe, parietal lobe, occipital lobe*, and *temporal lobe.*

- **Frontal Lobe-** associated with reasoning, planning, parts of speech, movement, emotions, and problem solving
- **Parietal Lobe-** associated with movement, orientation, recognition, perception of stimuli
- **Occipital Lobe-** associated with visual processing
- **Temporal Lobe-** associated with perception and recognition of auditory stimuli, memory, and speech

Keep in mind that the brain functions as a vast and complex network, with all parts connected and working with each other. "Movement" may be associated with the parietal lobe, but many parts of the brain are involved with movement. Saying that a function is contained in only one part of the brain is both simplistic and wrong.

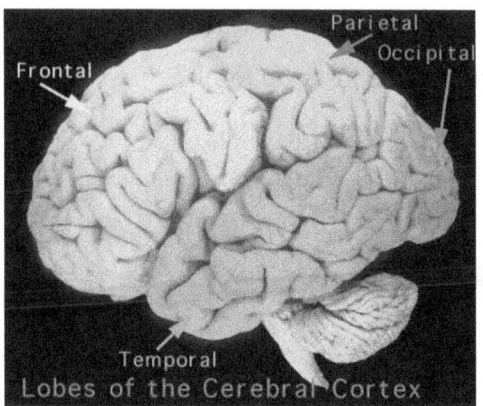

Figure 4-6. Lobes of the Cerebral Cortex
From Bryn Mawr College

There's also the *brainstem*, shown in Figure 4-6 at the bottom of the brain on the right, and is the oldest part of the brain. The brainstem controls basic life support systems (such as breathing and heartbeat), the activity of hormones, and primal emotions. Clearly, the brainstem is involved with the biochemistry of consciousness rather than the essence. The entire brains of many animals resemble our brain stem. Those brains without more complexity would not be able to link to an essence.

Essence and the Brain

There's a separate but related part of the cerebral cortex called the *limbic system*. This area is also known as the "emotional brain" because it controls most of the involuntary aspects of emotional behavior related to survival, such as anger, fear, pleasure, pain, and affection. Consciousness gets its direction from the limbic system and some aspects of the essence are associated with it.

But the cerebral cortex is only the outside ¼ inch of the brain. There are important parts of the brain under this covering, some of which you can see in Figure 4-7. From an essence perspective, we'll focus on those areas involved with your personality: the *cerebellum*, the *amygdala*, and the *hippocampus*.

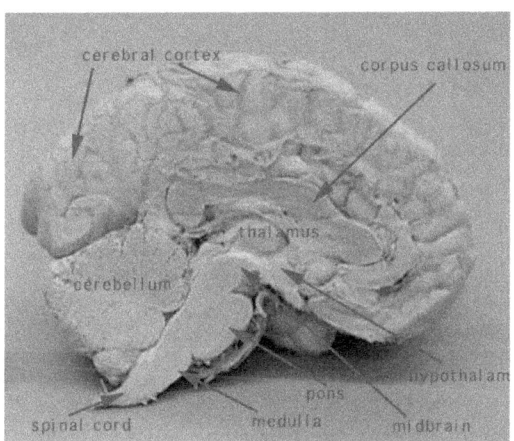

Figure 4-7. Human Brain Structure
From Bryn Mawr College

- **Cerebellum** – The cerebellum controls your movement, balance, posture, and coordination, and is also linked it to thinking, novelty, and emotions. *Cerebellum* comes from Latin meaning "little brain."
- **Amygdala** – The amygdala helps control your emotions. Be happy you have it, because without it you couldn't be happy.
- **Hippocampus** – The hippocampus forms and stores memories and is involved in learning. Without a hippocampus you couldn't remember anything, and people afflicted with Alzheimer's have lost the functioning of their hippocampus.

Human Brain Functions

Figure 4-8 shows some functional areas that have been identified. The frontal lobe, specifically the prefrontal cortex that controls aspects of personality, is the area most associated with the essence. Brain processing is so distributed, however, that other areas may also have important interactions. Joaquín Fuster notes that the occipital, parietal, and temporal lobes largely support perception and perceptual memory, while the frontal lobe supports action and executive memory. Eben Harrell reports that consciousness is now considered to be localized in a network consisting of a section of the prefrontal cortex and a section of the parietal cortex, with the thalamus controlling the signals between them. If any of these network connections is severed, consciousness ends.

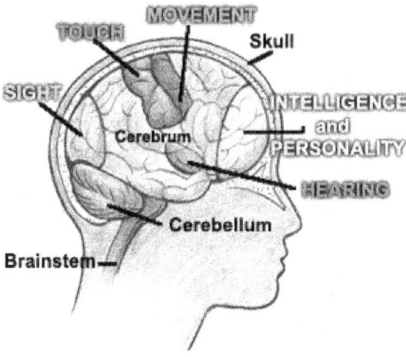

Figure 4-8. Specific Human Brain Functions
From Child and Youth Health, Adelaide, Australia

You might wonder how we know which parts of the brain do what. Most of the information about brain area functions has come from animal studies and from people with brain injuries, in which a focused part of their brain that was damaged resulted in changes to the person. For example, Phineas Gage was a railroad construction worker in the 1860s when an explosion caused an iron rod to pierce his frontal lobe (Figure 4-9). His personality changed to became a fitful, irreverent, indulgent individual prone to the grossest profanities. Similarly, personality changes have been noted in soldiers with frontal lobe injuries. As more information became available, the prefrontal cortex was identified as that part of the brain that controls personality.

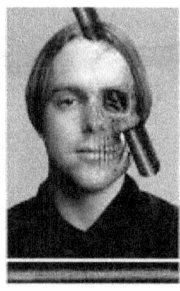

Figure 4-9. Phineas Gage

Gage's personality change was dramatically shown through the partial destruction of his prefrontal cortex, in which the balance between his basic animalistic tendencies (from older parts of the brain) and the civilized conduct he had learned (from the frontal cortex) was gone. Another perspective is that Gage's injury reduced the link with the Essence Dimension, resulting in an attenuated personality that wasn't able to control the animalistic tendencies that are natural in carnivores.

Essence and the Brain

Weak interconnections between the frontal cortex and the rest of the brain have also been observed in criminals, sociopaths, drug addicts, and schizophrenics. In these cases the Essence Dimension link is intact but neural connections in the brain inhibit personality control over the physical body. Not having the essence able to fully perform its executive management functions could result in frustrations leading to psychoses common in such individuals.

The prefrontal cortex in hominoids today is six times bigger than its size five million years ago, although total brain size is only three times bigger. Fuster notes that the prefrontal cortex is the most richly connected of all the cortical regions. Clearly, it's an increasingly important part of the brain, but its position in the skull also makes it very prone to injury. Wear your helmet when you bike.

Injuries to parts of the brain have allowed us to map physical body functions controlled by those parts. Can we do the same thing for the essence? Since the essence is reflected in an individual's personality and the area of the brain that controls personality is the prefrontal cortex, we may be able to isolate individual personality traits when more sophisticated brain analysis techniques become available. Some hints about where essence activities happen are available now, though.

Other important brain functions are memory and the mind. Let's look at how they work.

Memory

While most of memory is in the brain's larger lobes, Fuster points out that the prefrontal cortex contains a form called *executive memory*. This area seems to be important in higher-level functions such as planning, but we haven't established what memories are actually stored there. I speculate that our personality is initially moved from the essence dimension to this executive memory.

The Mind

When you think of your mind, you associate it with your brain, right? Not everyone agrees with that. Candace Pert points out that some researchers take a full body, psychosomatic networking view of the mind, with awareness being a product of the whole body and the brain functioning as one node. Other nodes include the spinal cord and the sensory organs. Anyone with a teenage boy suspects that his mind is often not in his brain.

The mind appears to be part of both the physical body and the essence. Physical aspects (chemicals, etc.) of it are part of the physical body; ethereal aspects of it are part of the essence. The essence is expressed by the individual's personality; aspects of the personality direct the formation of brain chemicals and some brain chemicals can affect attributes of the personality. While such complex interactions seem to make the physical body/essence connection in the mind impossible to separate, recall that if you can touch it, it's part of the physical body. Just keep in mind (so to speak) that all the chemicals are part of the physical body while some of the ethereal elements are part of the essence.

Some of those included ethereal elements are the *id*, the *ego*, and the *super-ego*. Freud said:

> We assume that mental life is the function of an apparatus to which we ascribe the characteristics of being extended in space and of being made up of several portions [i.e. the id, ego, and super-ego]*

---

* part of the unfinished book, "An Outline of Psychoanalysis" by Sigmund Freud. It was published in 1940, three months after Freud died.

Your Essence, Your Eternity

The id has been called the "pleasure principle" because it's an unorganized part of your personality that contains your instinctive drives and passions (such as food, water, and sex) and that seeks to avoid pain. It's amoral, doesn't take "no" for an answer, and acts generally infantile. Fortunately for you and everyone around you, your ego helps to mediate your id.

The ego is the reality part of your personality that helps to satisfy your instinctive drives in beneficial ways. It includes reason and common sense and generally seeks to ensure your safety by encouraging your adherence to laws, morals, and customary taboos. Your conscience is, in part, expressed by your ego.

The super-ego is the expression of your goals, including your perfection at the tasks you undertake. The policeman part of your conscience is expressed by your super-ego, which includes your understanding of right and wrong and punishes your misbehavior with feelings of guilt.

Complex interactions occur with the interplay of the id, ego, and super-ego, making analysis and adjustment of personality factors difficult and psychiatrists rich. To add to this complexity, people who participate in deep meditation may achieve a loss of spatial boundaries in which the individual sense of "self" is lost while a sense of global "oneness" is felt.

Similarly, while the mind sounds like the place we'll find the essence, we can't be sure until more research is done. Since the essence link is in the brain, you may say, where else could it be? At this time there isn't another place, but evidence must be found before this conjecture can be considered science. What we do know is that the mind is the place where intellectual activity is done and every living thing with a complex brain has a mind associated with it.

Are human minds different than those of other animals? Darwin argued that the human mind is part of the continuity of minds across animal species, and most subsequent researchers have supported that position. Some researchers, however, suggest that human minds are fundamentally different, and Marc Hauser points out four unique aspects of human cognition that are not exhibited by other animals:

1. **Generative computation**. This ability permits humans to form a virtually limitless variety of words, concepts, sequences of notes, strings of mathematical symbols, and other expressions. New ideas can be generated by combining old ones.

2. **Promiscuous combination of ideas**. If you stop giggling long enough to look up "promiscuous" you'll see it has to do with the indiscriminate mingling of things. In this case, it means people routinely combine ideas from different domains of knowledge, such as art, space, friendship, and (yes) sex to yield new approaches to social relationships, technologies, or laws.
3. **Mental symbols**. Humans can spontaneously express any experience in symbolic form that we express in language, art, computer code, music, etc.
4. **Abstract thought**. As far as we know, only humans visualize abstract thoughts such as the essence.

What's your opinion? As long as those four unique aspects result only from biochemical actions, the suggested superiority of the human mind is not of interest to us because intellectual abilities are part of the physical body. However, any non-physical constituent of those aspects would imply a different essence for humans from that of other animals with complex brains, and such a difference is contrary to our initial suppositions.

Essence and the Brain

Brain Development

How does all this happen? When a human baby is only four weeks into gestation the brain is developing at a rate of 250,000 cells a minute. Eventually there will be billions of neurons connected using perhaps a thousand trillion connections, with every cell and every connection precisely organized. Nothing is random and nothing is arbitrary.

By the time a child reaches two years old, its brain is about 80% the size of an adult brain. By age six, it's 95% of adult size, but the cerebral cortex continues to thicken and make more connections throughout childhood. The prefrontal cortex, where decisions are made, is in an early stage of maturity. Damage to it, as happened to Gage, destroys the physical connections that make maturity possible.

By 12 years old the thickening has peaked and excess connections start to be pruned. What's an "excess connection"? One that isn't used. So, by the age of 12, individuals can solidify capabilities within their brains by simply practicing them. It's the "use it or lose it" principle. For example, Thomas Elbert and colleagues found that string instrument players (banjo, etc.) who started playing before they were 12 years old had stronger signals from their brains for their left hands than did players who started after that age. Kazuko Eguchi is able to teach people to have perfect pitch (the ability to identify any musical note simply by hearing it), but you have to receive this training before you're 4 years old for your brain to be fully "wired" for it. It's much harder to develop a capability later in life, when the connections have to be re-connected.

One part of the brain that hasn't fully developed by the age of 12 is the prefrontal cortex. That's one reason teenagers are so loopy during adolescence – the judgment area of their brain is still under construction. Just because a teenager is physically mature doesn't mean s/he will reason in a mature way. As Figure 4-10 shows, the brain areas used by teenagers and adults when recognizing emotions in others are different. Teenagers rely more on the lower portion of the brain used for emotional purposes rather than the reasoning analytical area in the frontal lobe that's used by adults. Jay Giedd and Deborah Yurglun-Todd helped to understand these differences.

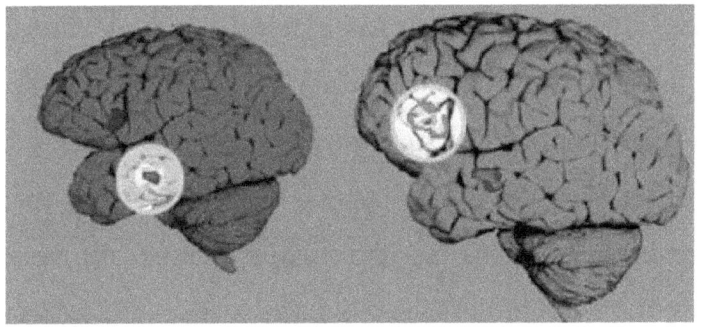

Figure 4-10. Emotional Analysis Areas
(From Yurgelun-Todd)

Bennett & Baird point out that physical maturation changes to the brain continue past adolescence and into the twenties. Brain development has historically focused on the *gray matter*, that stuff between your ears where mental computations happen and memories are stored. The *white matter*, which fills nearly half your brain, is the neuron connections that are coated with a white fatty substance called *myelin*. Gray matter contains neural cell bodies that store information while white matter is mostly myelinated axon tracts with no neural cell bodies. Myelin has been considered to be a kind of insulation for the nerve cables and has basically been ignored. Research since 2005, however, shows that the myelin wrapping forms in stages from birth to about 30 years and is involved in learning, self-control, mental illnesses, and even pathological lying. Myelination generally starts at the back of the cerebral cortex (near the shirt collar line) and is completed in the frontal lobes, where higher-level reasoning, planning, and judgment occur. There are suggestions that the smaller amount of myelin in the frontal lobe during the teenage years is one reason these kids sometimes make awful decisions.

The prefrontal cortex is part of the frontal lobe and is the most recently developed part of the brain. It's also the one in which most higher-level activities are processed and is the last area to complete the myelination process. At birth the prefrontal cortex is almost completely gray matter, with myelin sheaths basically absent. Dennis Carmody and his colleagues found that myelination occurs the fastest during the first three years of life, and Peter Huttenlocher notes that neuronal density is maximal at birth and declines by almost 50% (to adult levels) by the time the child is 7-10 years old, at a rate that correlates with the myelination process. What happens is that gray matter (neural bodies that store information) is connected with white matter (the myelinated tissue through which messages pass between different areas of gray matter), allowing those neural bodies to send and receive information efficiently. Why are we talking so much about myelination? Because it has significant implications for the essence that are discussed in Chapter 5.

After the myelin coating has been completed the brain stabilizes, continuing to build connections when focused on skill-based activities and continuing to prune unused connections. The next noticeable difference occurs as old age approaches, generally around 60 years old. Most people have a decrease in brain weight, probably due to loss of neurons, and a decrease in size of many areas of the cerebral cortex.

Essence in the Brain

Here's what seems to be the brain's role for your essence:

- Your prefrontal cortex is the link for your essence because the neural density and associated connections are both greater in the prefrontal cortex than elsewhere, and an essence may require a certain neural density to function. Think of neurons in the brain like water vapor in the atmosphere. As more water accumulates in the atmosphere eventually something totally unexpected is produced – rain. In the brain neurons collect in sufficient complexity to support an essence. Biological development has been described in a similar way – an increasingly complex aggregation of molecules collects to form genetic structures.

- Your mind contains aspects of both your physical body and your essence. The physical parts of your mind (chemicals that cause behavioral and other changes) are part of your physical body. The ethereal parts of your mind (personality, emotions, etc.) are part of your essence.
- Your essence is most directly seen as your personality. Contemporary psychological research uses the Big Five Personality Factor domains of Openness, Conscientiousness, Extroversion, Agreeableness, and Neuroticism to describe personality traits. Colin DeYoung and his colleagues found physical body differences in the brains of people expressing attributes associated with the Big Five. For example, self-described extroverts had significantly larger medial orbitofrontal cortexes, the area just above and behind the eye socket that's involved with keeping track of rewards. People with many characteristics associated with agreeableness

had larger posterior cingulated cortex and superior temporal sulcus areas, which are associated with understanding the actions and mental states of others. Above-average conscientiousness was associated with a larger lateral prefrontal cortex, which is involved in planning and voluntary control of behavior. Self-described neurotics had a significantly larger mid-cingulate gyrus, a region that detects pain and error, along with smaller brain volumes in areas known to regulate emotion. The researchers didn't find any differences in neuroanatomy for openness, which reflects imagination, curiosity, intellect, and artistic interests. Such findings illustrate the close and complex relationship between behavior (physical body) and personality (essence). Using aspects of the Big Five Personality Factors to improve your essence is explained in Chapter 10.

Your Exclusive Essence

Let me confirm what you already knew – you're a very special person. Unique, in fact. Your essence, exhibited as your personality, is exclusively yours. Of course, the same can be said for every living thing able to contain an essence, which includes all mammals and birds. While this may deflate your ego a bit, you're still unique!

Saying that every essence (personality) is unique is like saying every snowflake is unique. They *seem* different, but proving it is impossibly hard. Besides, your essence is in another dimension, with only a link by your body to access it. How can we know they're all unique? And, as has been pointed out several times, personality (essence) is strongly associated with behavior (physical body), and then you add culture and habitat to the mix to finally generate how an individual acts. With all of that, every person does seem unique.

We might get a better understanding of the essence link with the brain by looking at how we communicate.

Essence and the Brain

Chapter 5
Communication

There are four kinds of communication involved with a living thing. The first is that contained in the genetic code, expressed through automatic actions by the individual; the second is what a living thing learns as it grows; the third is directed communications between individual living things; and the fourth is the collection of past and current experiences of the essence. Let's look at each.

Genetic Communication

The first kind of communication is contained in the genetic code. We won't discuss genetic communication directed only to the individual (such as baby turtles, born on a beach, who automatically head for water as soon as they hatch), but will restrict ourselves to genetic communication expressed by an individual towards situations it encounters. Examples include fear and protection of loved ones.

We sometimes see fearless individuals or individuals who purposely put themselves in harm's way to protect a loved one. Fearless individuals do have innate fear but suppress it through instinctive reaction, determination, ignorance, or when interacting with others as a show of strength. Another reason an individual may be fearless is if that individual is an infant. Infants, for some reason, don't experience fear.

An example of heroism is Ginny, a Schnauzer-Siberian husky dog who rescued cats. She once threw herself against a vertical pipe at a construction site to topple it and reveal the kittens trapped inside. Another time she ignored the cuts on her paws as she dug through a box of broken glass to find an injured cat inside. The Westchester Cat Show named her Cat of the Year in 1998. When Ginny died of old age in 2005, 300 cats attended her memorial service.

A good description of basic fear comes from Audie Murphy, the most decorated American soldier of World War II who received 33 awards, including five from France, one from Belgium and the highest U.S. award, the Congressional Medal of Honor. Murphy said, "I was scared before every battle. That old instinct of self-preservation is a pretty basic thing, but while the action was going on some part of my mind shut off and my training and discipline took over. I did what I had to do."

Communication

When a gorilla averts its eyes, it's communicating its submission; when a bee dances, it's communicating the location of food; when a plant releases pheromones in response to an insect attack, it's communicating to other plants to prepare their defenses. While genetic communication is part of the physical body, it can be modified by training or through adjustment by the essence.

Communication Through Learning

The second kind of communication is what a living thing learns as it grows. Tiger cubs expand their innate playfulness into hunting knowledge by learning from their parents. Ants learn the situation outside of their nest (food available, defense needed, etc.) by touching antenna and watching the actions of knowledgeable ants. Humans have broadened learning communication beyond what's needed for survival into forming the life direction for the individual. Formal and working educations direct us to our careers. Social learning teaches us how to interact with our fellow man (sometimes by teaching us prejudice, a trait we're not born with). We learn all the good and bad attributes of life through learning communication − selflessness and selfishness, love and hate, pride and guilt, etc. Some of learning communication is part of genetic communication, but the knowledge extent is vastly enhanced.

So much of human communication is verbal communication that those individuals with speaking problems can have real difficulties in life. The two most common speaking problems are fear of public speaking and stuttering, both of which are almost always learned.

Public speaking is one of the most horrifying situations in life for some people, including those who have no trouble speaking to individuals, and those people will do almost anything to avoid it. Some studies put fear of public speaking in first place, ahead of the fear of death. Other statistics are that 3 out of every 4 individuals suffer from speech anxiety and up to 5% of the world's population aged 18-54 have a speaking phobia. And, of course, with the importance of giving presentations as a part of many jobs, a fear of public speaking can have a distinctly negative effect on your career.

Directed Communication

The third kind of communication is directed communications between individual living things. Interspecies communications occur during close interactions, such as people talking to pets (or the other way around), but the quality and quantity of such communications is so limited that little is known about the true recognition of either party. Within a species information exchanges almost certainly occur, but little research has been done to describe it. It's known that dolphins have a signature whistle to identify themselves and use a complicated system of whistles, squeaks, moans, trills and clicks produced by sphincter muscles within their blow hole, but what these sounds mean to other dolphins is unknown. Pet owners have classified various sounds and body language used by their non-human friends, but an understanding of what they mean has not been determined. With the limited amount of verbal expressions and the obvious need to communicate between individuals, it seems that animals must use forms of non-verbal communication that have yet to be understood by humans (except in stories like Doctor Doolittle). Extensive communications research has been restricted to human communications. With all the natural ways humans communicate (rich language features, pitch and volume sound variations, body language) and technological ways we communicate (telephone, television, written material, telecommunications), it's astonishing that we have so much miss-communication.

The more you know a person the more you understand their communications. There are plenty of examples of people who have been together for many years who can anticipate what the other person will say or do before they say or do it. In these cases, miss-communication occurs when one of the people changes the way they act. Physical body problems, such as hormone changes, accidents, or disease may be the cause. If you don't have such issues, just being yourself will keep your communication lines intact. If you continue to act oddly your close friends may reconsider the kind of person you really are. Of course, if you were a jerk to start with, personality changes may result in a distinct improvement in your personality and improve your essence.

Communication

Essence Communication

The fourth kind of communication is the collection of past experiences of your essence. Perhaps you've had experiences like those of H.G. Wells, who wrote that at times in the night and in rare lonely moments, "I experience a sort of communion of myself with something great that is not myself."

Although essences reside in another dimension, communication with them may be possible. A way to illustrate this is by an example from *Flatland*, a novel written in 1884 by British clergyman Edwin A. Abbott. Flatland is a 2-dimensional world inhabited by squares, rectangles, etc., like drawings on a piece of paper. These Flatland inhabitants can only communicate in two dimensions, so when a 3-dimensional person draws a new object, the Flatlanders can't see upwards (the third dimension), but only see the new object being magically formed. They have no way of knowing where it came from or how it got there.

Let's extend this example to the house of a 2-dimensional Flatlander (Figure 5-1). The larger openings are where inhabitants can move into the house and go from room to room. The smaller openings on the outside are windows. Note that since the Flatlanders are 2-dimensional they have to move through the openings to get into and around the house, but that we, as 3-dimensional beings, can see everything inside by looking down our third dimension.

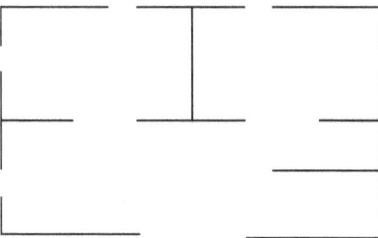

Figure 5-1. Flatland House

Moving this analogy up a dimension, how would we interact with beings in higher or non-spatial dimensions? We have no knowledge of their languages or any way they communicate with each other. But those other-dimensional beings would be able to communicate with us, because everything we do is visible to them in the same way that everything in Flatland is visible to us. In this case, "them" are essences.

There have been many reports about communicating with deceased persons and quite an industry has existed since ancient times to help you do it. It's not unusual for close family members to feel the "presence" of a deceased relative for quite a while after the person has died. This could, of course, be an extension of longing for that person, with your feelings amplified so much that you imagine the person is still hovering around nearby. Such feelings provide a measure of comfort. But there's another possibility, too. While children claiming to be reincarnated don't generally focus on the previous person's funeral, the few statements that have been made and collected by Stevenson imply that the consciousness of the previous person remained near the body or the family for some time after death.

What's the extent of this communication? People have reported strong feelings of the departed person's presence or a desire to go (or not go) to certain places. When someone says physical objects were moved ("Those dishes flew straight across the room!") they're not talking about essences. Only physical bodies can move physical objects. If the person contradicts with "I know a ghost was involved" they're still not talking about essences. An essence is linked in the brain of a physical body and is reflected in the personality of that physical body. Essences don't float around scaring people or throwing dishes. What you will get is the real feeling that the departed person was sending some guidance.

Historical Experiences

We've mentioned how your essence will have had historical experiences it remembers from previous physical bodies. Jim Tucker describes his research with young children in this area by noting:

> Emotions, attachments, fears, addictions, likes and dislikes, and even identification with a particular country and with a gender may be able to carry over from one life to the next. If reincarnation does occur, emotions as well as memories survive.
>
> The emotions do not necessarily continue throughout this life. The behaviors often persist past the point when the children stop talking about a previous life, but they generally fade away over time.

What about communicating those previous life memories? As we saw in Chapter 4, myelination proceeds from the back of the brain forward and new information first replaces old (and unused) memories in the hindbrain. Memories stored in the prefrontal cortex have the best chance of being retained long enough to be expressed by the child. Peter Huttenlocher reports that neuronal density in the prefrontal cortex is maximal at birth and declines by almost 50% by the child's 10th birthday; a child's language skills become proficient enough for adult communication when the child is three or four years old; and Stevenson reports that memories of a past life start to fade by the time the child is eight years old. Perhaps memories from a previous life are initially stored in the prefrontal cortex and are recalled by the child as s/he learns to talk, but those memories disappear as the neuronal density declines.

But how did the past-life memories get in the prefrontal cortex gray matter in the first place? My conjecture is that these memories are projected from the Essence Dimension when the link was first established and stored in the executive memory of the prefrontal cortex, along with basic information needed for initial personality development.

Essence Communication with the Brain

Interactions between the essence and the brain must be done with some kind of communication method. What can it be?

It's tempting to say the method uses the brain's Beta, Alpha, Theta, or Delta Waves, but it doesn't. Those waves determine brain activity by measuring the intensity and frequency of electrical pulses generated by the synapses. Even if you were a very simple animal, like a flatworm, an appropriately sensitive EEG could measure the synapse firings and the generated waves. No, essence communication is only done by complex brains, and a communication method specific to those brains must be used to achieve this ultimate level of communication. One possibility, involving a relatively new way physicists suggest to describe how everything in the universe works, is explained in the next chapter.

Chapter 6
Essence In String Theory

Some physicists have tried to use their science to understand personality, emotions, and other features of the essence. The three dimensions of space (length, width, and height) and one dimension of time that we're familiar with are inadequate to explain non-physical events, but Rauscher & Targ point out that a geometrical model of spacetime called *complex Minkowski space* may explain them. The mathematician Hermann Minkowski was an early believer that time is a fourth dimension. Einstein, a contemporary of Minkowski, picked up on this and the mathematical setting used to describe Einstein's Theory of Special Relativity is called *Minkowski Space*. Some mathematical descriptions include a "complex" component, which means answers in irrational numbers (such as the square root of minus 1). Perhaps there's also a complex explanation for some non-physical parts of life.

Your Essence, Your Eternity

Stanford University physicist Andrei Linde says, "I cannot imagine a consistent theory of everything that ignores consciousness." The "Theory Of Everything" (TOE) he's talking about has sometimes been called the *Grand Unified Theory* (GUT), which includes gravity, relativity, and quantum mechanics. You can choose which body part to use as the acronym – physicists will accept either. The idea is to describe interactions within the universe at both the large scale (such as stars) and the small scale (such as photons) within a single theory. A GUT/TOE was the goal that Einstein worked on for most of his life but was never able to complete. It's the most sought-after goal in theoretical physics. In this view, the essence is an independent force that must be included with and consistent with the other forces of nature.

Physicists developed *String Theory* to better understand curious events they see all around us. I mean all around us, from where you're sitting to the end of the universe; from the tiniest subparticle no one can see to the largest supercluster of galaxies our telescopes observe. String Theory focuses on matter and energy, but we'll only cover enough of this big subject to help us understand how concepts in it provide the framework for the essence features of personality, attitudes, morals, and ethical evaluations.

Background

In 1687 Sir Isaac Newton published his Laws of Motion, which explain how gravity works and how stars and planets move the way they do. Newton's Laws said that light always travels in a straight line, but in 1915 Albert Einstein speculated in his Theory of General Relativity that light curves when it's close to a massive object. The total solar eclipse in 1919 proved that Einstein was correct when light passing close to the sun showed a star that was actually behind the sun. You might think such a discovery would make him famous, but he was already famous in 1905 for his equation $E = mc^2$, part of his Theory of Special Relativity (1905a) to describe the relationship of energy to mass. He got the Nobel Prize for discovering the photoelectric effect (1905b), where light bounces off certain materials, making our automatic door openers possible. Quite a guy – the only scientist to have had a ticker-tape parade in New York City.

The Theory of General Relativity has been proven correct over and over for the past hundred years, but it's restricted to larger masses.

At the same time that Einstein was working on relativity, Max Planck and other outstanding physicists were looking at what happens at the smallest places with the many sub-atomic particles. They came up with *quantum mechanics* to describe what they saw (or tried to see). Richard Feynman, one of the leading quantum scientists remarked in 1965, "I think I can safely say that nobody understands quantum mechanics." Quantum mechanics has also been proven correct over and over for the past hundred years but is not compatible with the Theory of General Relativity. Einstein's equations fail at the smallest levels and quantum mechanics equations fail when measuring big things.

Physicists hate having two different theories to describe one universe, so they've worked very very hard to bring them together as the *Standard Model*. One thing they haven't included is gravity, and you can't have a theory of everything without gravity.

Essence In String Theory

Einstein uses points as the smallest entity to make big things. Quantum mechanics uses tiny moving bundles of energy, called quanta, to make small things.

String Theory, originally proposed by Edward Witten and that continues to be improved by him and others, uses tiny vibrating strings as the smallest entity for all matter and energy, big and small. Although many physicists hate String Theory because the strings are too small to test and the math is almost impossibly hard, it includes all the features of the Theory of General Relativity and quantum mechanics, plus gravity. We'll look at the main features of String Theory, starting with its dimensions.

String Theory Dimensions

String Theory has a maximum of 11 dimensions, with our universe including all of them. What are they? We'll start with the ones we know well.

Our senses help us understand our world, to measure and manipulate what's around us. We can measure and manipulate the length, width, and height of components in our 3-dimensional world. We can measure (but not manipulate) time, which we consider a fourth dimension.

Your Essence, Your Eternity

Our essence is also around us, but we can't measure or manipulate it because it's in a dimension that's not part of our world. The concept of what a dimension is may be confusing, so we'll take a moment to hopefully remove a little of the dim from dimension.

Although you use it in casual conversation, you probably didn't know that the word "dimension" is really a mathematical concept for defining objects. We'll ignore the math but keep the objects. Figure 6-1 shows the dimensions you already know.

Dimensions	Object	Description
0	.	Point
1	 •——•	Line (length)
2	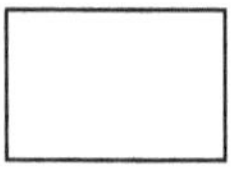 ▭	Rectangle (length & width)
3	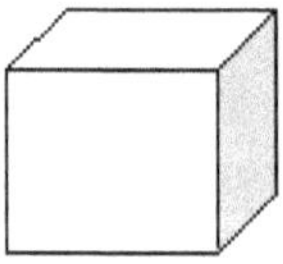 ⬛	Cube (length, width, & height)

Figure 6-1. Spatial Dimensions

Essence In String Theory

You know that you're a 3-dimensional object and anyone who calls you 1-dimensional is insulting you. All of these are *spatial* dimensions because they occur in space. Since String Theory also uses strings to describe energy, there are also force dimensions.

The description for dimensions seems a bit technical, and it is, because mathematicians and physicists use dimensional theory in constructing their views of the universe. Where are these dimensions? Except for length, width, height, and time, none of the other dimensions have been formally described. While conventional physicists throw up their hands at this, String Theory physicists consider themselves as guides to bring those disbelievers to the real truth.

The essence is considered a force containing information, so the Essence Dimension is a mix of the two. While a force projected from another dimension sounds like a radical idea, it's not a new one.

If physical forces and particles can be described by projections from outside of our universe, why not life forces? Consider for a moment that one of those dimensions in String Theory could be a life dimension for the essence (Figure 6-2).

Your Essence, Your Eternity

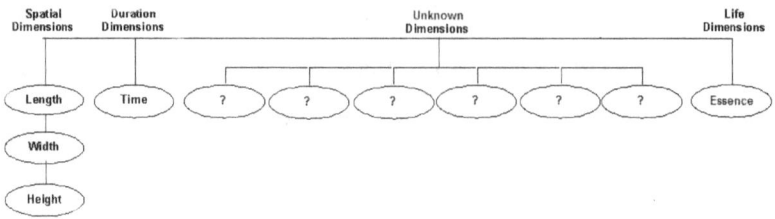

Figure 6-2. Essence Dimension

The essence contains those attributes you see as your personality. Our senses don't extend beyond three dimensions, but our prefrontal cortex communicates with the Essence Dimension instantly and continuously.

This is problematic because we don't even have a way of measuring the essence. How would this Essence Dimension interact with the prefrontal cortex in the brains of living things? Where would the essence be? What would the essence be doing?

The Essence Dimension

Note that although it's in another dimension, the essence is all around us all the time, just like gravity is always around us. An essence becomes evident when a new living thing with a container (brain) that can link to it is born.

The concept of separation of mind and body goes back to ancient times, with its philosophy established by René Descartes in the 17th century and updated by William James and Henri Bergson in the early 20th century. Called *substance dualism*, the idea is that the brain acts as a filter for stimuli reaching it. Referring to substance dualism, Ian Stevenson says:

> I believe that we are obliged to imagine a mental space that, necessarily, differs from the physical space with which we are ordinarily familiar. (p. 181)

Religions have attempted to describe such a separate dimension. For example, the Zen master Huang Po describes the ultimate state of being as:

> All the Buddhas and all sentient beings are nothing but One Mind, beside which nothing exists. This Mind, which is without beginning, is unborn and indestructible. It is not green or yellow, and has neither form nor appearance, it does not belong to the categories of things which exist or do not exist, nor can it be thought of in terms of new or old. It is neither long nor short, big nor small, for it transcends all limits, measures, names, traces, and comparisons.

Your Essence, Your Eternity

Since essences, as projections from another dimension, are outside of the duration dimension (time) and the spatial dimensions (containing the physical bodies), they're timeless and immortal. Unfortunately, we don't yet know of a mediator for the essence and haven't even been able to measure it. But, as pollster George Gallup said, "Just because we don't yet know how to measure something, that doesn't mean it doesn't exist."

Care must be taken to avoid confusing personality with behavior. For example, since the 1980s research on reasons for aggressive behavior has focused on low levels of the brain enzyme monoamine oxidase A (MAO A). Individuals with naturally low MAO A are more aggressive, and sometimes the researchers report that those individuals have an aggressive personality. What they really have is aggressive behavior (part of the physical body), which might cause them to develop an aggressive personality (part of the essence). The link between behavior and personality is strong.

Each essence may have had historical experiences that it "remembers" from previous physical bodies. Those experiences can make future affiliations with highly evolved physical bodies easier to handle.

In addition, the qualities expressed by the essence depend on the physical body, which is different for each kind of living thing and are modified by the habitat and culture of the living thing. In other words, humans and hounds have essences, but the expression of essence qualities is different. Simpler brains only permit the expression of more basic aspects of the essence. More complex living things have more "room" in their brains for essence elements. In this case "room" doesn't imply physical size, but refers to the capability of the brain to contain more aspects of the essence. Using a human example, let's see how an essence can become part of a person's life.

The Essence Process

An essence is linked from its dimension into a physical body as soon as brain development has progressed to include a functioning prefrontal cortex for personality development (Figure 6-3). This is a two-way transmission link from the essence to the brain and back again. Previous life information (if any) is transmitted to the prefrontal cortex.

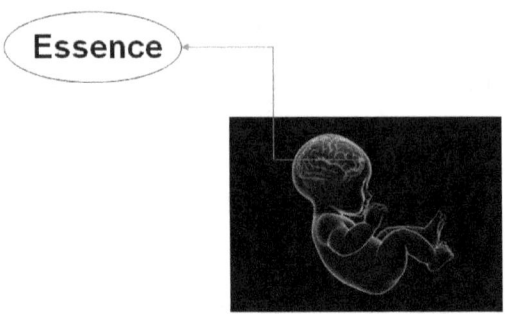

Figure 6-3. Initial Essence Link

As the brain continues the myelination process, at least some of that previous life information will be destroyed. By the age of 10, the prefrontal cortex working memory becomes fully occupied with daily activities and the previous life information is overwritten. That may be why the children Stevenson found who talked about being someone else before were always younger. But, if it's important to the person, that information could be saved elsewhere in the brain. That essence is associated with the living thing as long as that physical body exists (Figure 6-4).

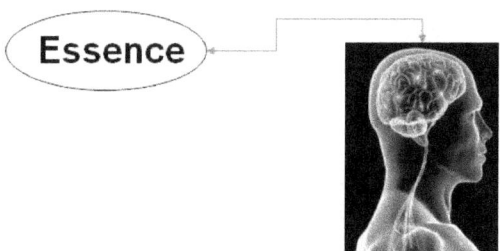

Figure 6-4. Brain-Essence Link Throughout Life

Throughout life, some current life information is transmitted back to the Essence dimension. When the physical body dies, the essence link is detached to await association with another living thing (Figure 6-5). Similarly, if a living thing doesn't completely die but can no longer link to the essence, that living thing is *brain dead* but remains alive because consciousness is still there but the essence isn't.

Figure 6-5. Brain Dead

The information sent to the Essence Dimension is preserved. At some future time, that same essence will be linked to a new individual and the process starts all over.

If you've gotten this far, you're curious enough about Stevenson's work to say to yourself, "If any of that reincarnation stuff is true, I want to improve my personality not just for my life now, but just in case there really is more to my future."

Let's see how we do that.

Part 2

Improving Your Essence

Your Essence, Your Eternity

Chapter 7 – Your Personality

Everyone wants to have a healthy, mature personality, with maybe a dash of occasional recklessness to make life interesting. Most of the research about personality has focused on behavior disorders or mental illness, with very little devoted to aspects of a healthy personality. Don't blame the researchers too much – nobody gives grants to study healthy people, only those who are a menace to society. The great majority of people with personalities acceptable to the culture are left to figure out how to improve themselves by themselves.

While the suggestions in this book will help you improve specific personality attributes, let's look at personality from a larger viewpoint first. We'll consider what it means to have a healthy personality from the perspectives of six giants in psychology – Gordon Allport, Carl Rogers, Erich Fromm, Abraham Maslow, Carl Jung, and Viktor Frankl – as summarized by Eddington & Shuman.

A Healthy Personality

Gordon Allport says our adult personalities look ahead to our futures rather than being pushed ahead by our past, which encourages a rather optimistic view of life. He says, "The possession of long-range goals, regarded as central to one's personal existence, distinguishes the human from the animal, the adult from the child, and in many cases the healthy personality from the sick." Being forward-looking increases your tension level, which Allport says is healthy. This is in direct conflict with Freud who believed that people were always trying to reduce tensions in their lives. Allport also thinks happiness is a by-product rather than a goal and that the life of a healthy personality can be painful and grim. Goals, in fact, are relatively unattainable – the healthy personality always pursues new ones and tries to perform at the highest level of competency possible. Does this sound like you? If so, here are Allport's criteria for a healthy personality:

- Extend yourself to be fully involved in work, family, and leisure activities
- Express unconditional love for parents, spouse, children, and close friends

Your Essence Development

- Express compassion for the troubles of others
- Accept yourself as is, with your weaknesses, and try to improve them when possible
- Tolerate frustration and try to find alternate ways to reach your goals
- Direct your emotions towards constructive channels
- Accept reality as it is without pretending it's something else
- Develop long range goals
- Develop values that unify your philosophy of life
- Include a sense of duty and responsibility as part of your conscience

Carl Rogers thinks that each person's perception of the present is the most important and he invented *client-centered therapy*, in which much of the discovery of a person's troubles are found by the person, not by the therapist. People attending a Rogers-type session may wonder why they're paying a therapist so much when they're doing all the work. Rogers thinks that a healthy personality is "a direction, not a destination" and that our goals in life should be towards increased complexity to become the best we can be at whatever we want to try. He also thinks that a person has an innate desire to create, with the most important creation being him/her self.

Rogers believes that the most important requirement for a healthy personality is for the child to receive unconditional love from its mother or caregiver, no matter how terribly the child acts. Of course, children are to be told when they're doing something wrong, but the unconditional love they see makes the child feel guilty about doing something that's disapproved and the child won't do it again. Doing good things makes the child feel that much better. Does this sound like you? If so, here are Rogers' criteria for a healthy personality:

- Act as the person you are without pretending to be something you're not
- Experience a wide range of both positive and negative emotions without becoming defensive
- Live fully in every moment, growing new experiences from the current ones
- Behave in a way that feels right rather than behaving by intellectual thought
- Behave in a creative and spontaneous way with little concern for the opinions of others
- Feel a sense of personal power about your life

Erich Fromm thinks that it's up to society to adjust to the basic needs of the individual, not up to the individual to conform to society. In other words, a healthy culture makes healthy personalities in its individuals. He thinks our behaviors are not instinctive but are developed through a union between others, nature, and ourselves. Without a successful union we become more "animalistic" rather than more comfortable within our culture. Does this sound like you? If so, here are Fromm's criteria for a healthy personality:

- Avoid becoming submissive to another person, group, or religion
- Avoid forcing others to submit to your will
- Develop a sense of integrity and individuality through love and solidarity
- Be an active shaper of your life by creating such things as children, ideas, and material goods
- Establish involvement, concern, and participation with society and other people
- Develop a realistic and objective view of the world without distortion by fear
- Attain happiness through productivity
- Use your conscience to self-direct your moral code
- Encourage political and economic systems to foster human personality growth

Abraham Maslow thinks that each person has universal needs that they need to grow and develop to the fullest, but that less than one percent of the people are able to do it. He thinks that childhood experiences can inhibit personality growth. Maslow described an optimistic but universal *hierarchy of needs* (see Figure 1-2) all of which must be at least partially satisfied, in order by importance as follows:

1. **Physiological needs**, including food, water, air, sleep, and sex.
2. **Safety needs**, including security, stability, protection, order, and freedom from fear and anxiety.
3. **Love and belonging needs**, including caring relationships and being part of a group or ideas that reflect our own.
4. **Esteem needs**, including recognition of others and self-esteem.
5. **Cognitive needs**, including understanding our world.
6. **Aesthetic needs**, including beauty, poetry, music, etc.
7. **Self-actualizing needs**, including developing our potential to become the best we can become.

If you think you're not functioning at your highest level, the hierarchy of needs presents a path to follow. Does this sound like you? If so, here are Maslow's criteria for a healthy personality:

- A good perception of reality
- A general acceptance of nature, others, and oneself
- Spontaneity, simplicity, and naturalness
- A focus on problems outside your own
- A need for privacy and independence
- Autonomous functioning

Your Essence Development

- A continued freshness of appreciation
- Peak experiences
- Social interest
- Interpersonal relations
- A democratic structure
- Discrimination between means and ends, and between good and evil
- A sense of humor
- Creativeness
- Resistance to being conditioned by cultural norms

Carl Jung thinks that unconscious experiences that we as individuals, that humans as a species, and that the animal ancestors of humans have accumulated are the most important. We must reconnect with the symbols, rituals, and myths of human history to understand those unconscious experiences, and then integrate them with our consciousness. He also described different types of consciousness and how individuals with them orient to the world. Jung identified a variety of universal experiences, the most important of which are:

- The mask we hide behind, and sometimes use to cope with undesirable events, when playing roles in life
- Masculine components in female personalities
- Feminine components in male personalities
- Deep emotions, including spontaneity, creativity, and insight, which can also be evil and destructive

- Our efforts to integrate all conscious and unconscious aspects of our personality into a whole

Does this sound like you? If so, here are Jung's criteria for a healthy personality:

- Becoming aware during middle age of the aspects of the self that have been neglected. Life's primary challenges (money, prestige, position) have been met and now the unconscious must be probed for new behaviors
- Rejecting the goals of earlier life in favor of those needed in later life
- Eliminating our masks to become who we really are
- Becoming aware of both the constructive and destructive aspects of our emotions
- Becoming comfortable as a man to exhibit some feminine traits and as a woman to express some masculine traits
- Develop a personality that combines all the universal experiences

Viktor Frankl thinks we have freedom to choose an attitude for any fate that may come our way, a perspective he developed from his own experiences in two Nazi concentration camps during World War II. He quotes Nietzsche with "He who has a *why* to live can bear with almost any *how*." To help people who feel their lives lack meaning, Frankl developed *logotherapy* based on the freedom of will, the will to meaning, and the meaning of life. Personal responsibility is the primary way to success. It's the opposite of being stuck with biological or any external force driving our lives. Does this sound like you? If so, here are Frankl's criteria for a healthy personality:

- Have a meaning to your life so there's a reason to continue living. Search for that meaning rather than for your self.
- Accept inner tension to see what you are and what you want to be
- Pursue creative and productive activities by producing products, generating ideas, or serving people
- Involve yourself in natural experiences in life, the more intense the better
- Have the freedom to choose your course of action
- Be personally responsible for what happens in your life
- Be oriented toward future goals
- Be committed to work
- Be able to give and receive love
- Accept your fate with dignity

The Golden Rule In Practice

How should those various "healthy" personalities interact with other people in real life? You can't go wrong with the most fundamental and universal principle in virtually every religion – the *Golden Rule*. The Golden Rule is

Do to others as you would have others do to you.

But what do you do when those other people aren't following the Golden Rule themselves? It turns out the most successful approach in dealing with people is the *tit-for-tat* strategy. This means that, in your first interaction with an individual, you should cooperate with them. Afterwards, you should do the same thing the other person does. If the other person cooperates, you also cooperate; if the other person does not cooperate, you don't cooperate the next time. Tit-for-tat punishes quickly and forgives quickly (you cooperate again when the other person does). This strategy has the advantage of starting out with the Golden Rule and continuing with it as long as the other person follows it. But if the other person violates the Rule you stop cooperating, which should show the other person that they would benefit more by cooperating with you. The Golden Rule would then be re-established for both people.

Quick response is a key element of the tit-for-tat strategy. However, very often individuals (and even companies and governments) don't respond immediately, with the reasoning that the transgression was small and there's no reason to punish the other person (or company or government) over such a small thing. What happens is that the offended individual stores this annoyance in the back of their mind. At a future time the same pair meets and the offender does something else small and annoying. Once again, the offended individual chooses not to make an issue of it, and the offender continues with no punishment. However, this offense is also stored. Then, one dark day, the offender does yet another small annoying thing and suddenly the offended person blows up, responding in a way that's far beyond what should be done for this small offense. The offender is completely surprised by such behavior, and now we have a major issue seemingly over a small offense. This is how friendships are ruined and (sometimes) how wars are started. It all can be avoided simply by quick responses to perceived offenses through the tit-for-tat strategy.

Dealing with all of life's ups and downs is a basic function of your essence. Let's see how to develop the best essence we can.

Your Essence, Your Eternity

Chapter 8 – Your Essence Development

The most basic essence elements become associated with you early in life, with more features added as you mature. This can be observed as a new baby, completely dependent on outside care for its survival, exhibits only the beginnings of its personality. As the baby grows, more personality attributes are observed and others become more fully developed. The full human essence comes after adolescence, when the individual matures into an adult with his/her personality attributes (attitudes, emotions, morals, etc.) in place.

Each person develops their personality qualities at their own rate, some slower and some faster than what's considered "average." This was expressed in the film classic *It's A Wonderful Life* in the following exchange:

> POP: He's pretty young for that job.
> GEORGE: Well, no younger than I was.
> POP: Maybe you were born older, George.

While people who act immaturely are troublesome at every stage, real problems occur when a person with an immature essence (reflected in his/her immature personality) reaches the legal age in a culture and has more independence to act immaturely. Let's review these stages of development and the personality qualities expected in them.

Stages of Development

Humans are considered newborn for their first month and infants for their first year. They begin a playful stage at this time and become more playful until they're about 8 years old, when they start to become more serious. From ages 1-3 they're considered toddlers. Children start preschool as young as 2 for reasons such as both parents have to work and preschool is the most convenient dumping ground for their kid, or the parents think their child is a genius and send him/her to a fancy school to polish their intellect, or the kid is driving his/her mother crazy and she needs some relief. Primary school starts when the child is 5 or 6 and continues to about age 12. From about ages 10-12 children are considered preteens and girls start to enter puberty, causing great consternation to their fathers but not affecting their male classmates until their own puberty starts at about age 12. This is both the adolescence stage and the time when children are in secondary school. They graduate at about age 18 and start their legal adulthood, but their psychological maturity doesn't necessarily follow this timetable.

Standard aspects of human development are physical growth, motor development, cognitive/intellectual development, social-emotional development, and language development. Social-emotional development is our main interest because that's where attributes of the personality (essence) are found. However, cognitive/intellectual development, which is really an attribute of the physical body, has such a close link with personality that we'll include it to complete and complement our discussion of essence association.

The most accepted research on cognitive/intellectual and social-emotional development was done by Jean Piaget who described the following four stages. These stages describe "normal" children, not children with developmental disabilities.

- **Sensorimotor stage** from birth to age 2 during which children experience the world through their movements and their five senses. During this stage children are extremely egocentric and are unable to see the world from any perspective other than their own.

- **Preoperational stage** from ages 2 to 7, during which their egotism wanes and they become more like people you want to keep around. This is when they start to learn empathy and social rules, which continues into adulthood.
- **Concrete operational stage** from ages 7 to 12 when children begin to think logically but they're very concrete in their thinking. For example, they might think that as long as they stay out of their bedroom it's not bedtime. Concrete thinkers are often quite stubborn and this age is a good time for moral stories about stubborn people (or animals) changing their minds for the better.
- **Formal operational stage** from age 12 children become better at logical thinking and start to develop abstract thinking.

Essence Maturation

How are personalities matured as we move through adolescence and into adulthood? Why do individuals exhibit different personality traits in their essence, even individuals from the same family? In other words, why are some people caring and thoughtful while others are selfish and thoughtless? The answer lies in the nature/nurture philosophy, but with a revised description of nature.

The significance of nature or nurture, the genetic heritage of an individual or his/her environmental upbringing, was debated around the year 400 by two pillars of the Roman Catholic Church, Pelagius and Augustine (later Saint Augustine). The question was whether Adam's original sin of biting the apple in the Garden of Eden was only his problem or was a problem for everyone forever afterwards. Pelagius argued for free will, saying Adam's disobedience ended with him. Augustine argued that Adam began a characteristic inherited by all humanity, that of an inherent tendency to commit sin. While Augustine's arguments were a clear reflection of his own well-known and uncontrolled lusts, his view won out and became the official position of the Church. Consequently, the commission of violence has been designated as Adam's legacy and can only be pardoned by a divine being. Clarence Darrow, the famous defense attorney, used the concept of inherited sin in his defense of Nathan Leopold and Richard Loeb, two young men who were admitted murderers trying to create the perfect crime. Part of Darrow's argument was, "this terrible crime was inherent in his organism, and it came from some ancestor." Darrow succeeded in his goal – Leopold and Loeb received life sentences rather than the death penalty.

Racist individuals have used any positive research about nature (or negative findings about nurture) to claim that genetic heritage is the key. Some of the most atrocious acts committed by humans have been done in the name of genetic cleansing. Sociologists, on the other hand, have tried to show that the quality of an individual's environment makes all the difference. Arguments for social improvement programs use the positive research on nurture to press for money to improve poor environments. Whenever the question of nature vs. nurture is raised, care must be used to ensure that extreme conditions, such as starvation or poor health due to poor health care, are excluded. Equally confusing would be extreme wealth, which could mask a terrible home environment.

Nurture has elements of both the essence and the culture components, while nature is split between the physical body and the habitat components. An individual's personality is part of their essence. Each essence is completely independent from the essence of another individual.

The independence of the essence from the physical body also shows how disparate personalities can occur in the same family. As a baby matures, s/he is guided by the parents and other people close to the family, an environment that encourages consistency among the individuals. Such consistency is more likely with an essence similar to that of family members, but the baby may have an essence so different that s/he doesn't conform to the group. In that case, the "black sheep" of the family follows his/her own path. Daily living permits little time to ponder some unusual things that are actually rather common. For example, why does your sister act so weird, so different from everyone else in the family? We say the "black sheep" follows her own path and we forget it and have another beer, but the question remains.

An interesting example (Figure 8-1) is Dr. William M. Bulger, former President of the Massachusetts State Senate and former President of the University of Massachusetts at Amherst, who's also the brother of James J. (Whitey) Bulger, a mobster who was on the FBI's 10 Most Wanted List. They must have had arresting family dinners.

Figure 8-1. William and James Bulger

How about identical twins, individuals conceived from the same egg that divided into two separate embryos (Figure 8-2)?

Figure 8-2. Identical Twins

Do they act the same? Their genetic makeup is the same and they usually act very much alike, but they have different personalities. Since no "personality" genes have been identified, published research on identical twins raised together suggest differences in their environment are the reasons for differences in their personalities.

It should be pointed out that researchers suggesting other, more "unusual", reasons find their papers aren't published in respected academic journals, their funding dries up, and their tenure opportunities vanish.

Identical twins are fine, but more convincing evidence will come from clones.

Clones

Clones, as they're usually understood, are duplicates of the physical body. Is their personality identical, too?

A pet and stunt performing Brahman bull named Chance (Figure 8-3) was incredibly docile and became the "business partner" of ex-rodeo clown Ralph Fisher, who took him to fairs and parades. Fisher decided he wanted such happy days to continue, and Chance was cloned as Second Chance.

Figure 8-3. Chance the Bull

Unfortunately, Second Chance wasn't a second Chance. As reported by ABC News on January 30, 2009 (*Cloned Pets: Looks Can Be Deceiving*), on his first birthday he threw Fisher up in the air and a few years later gored him with his horns. Cesare Galli, a prominent researcher with horse and cattle clones, told me "With cattle I can tell that they differ between them in their character" (personal correspondence).

To really study clones you need a laboratory setting, not a barnyard, and enough identical animals to provide data that's harder to question. The obvious choice is our old friend, the mouse. Using genetically identical mice living in the same environment, Gerd Kempermann, a behavioral geneticist at the Dresden University of Technology, wondered if they would develop different personalities that would also affect the brain development of their hippocampus. Putting 40 such animals in a mouse paradise filled with toys and environmental enrichment, his team found that the mice reacted differently, with some becoming explorers and some staying in one spot (Figure 8-4). Their report (Julia Freund and her colleagues) concluded in part that distinct personalities were developed despite their genetically identical bodies.

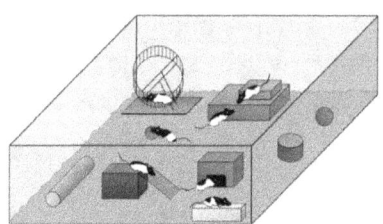

Figure 8-4. Cloned Mice in a Mouse Paradise

This is certainly not what's desired when creating a clone – you want the personality of that individual, not just the body package it's in. The separation of the personality from the physical body means that cloning your pet (or you) will only endow that clone with physical body traits, not personality traits. Cloning duplicates the genetic structure of an individual into another body, producing an exact copy of that original individual. But it's only an exact copy of its physical body, not its personality. So, if you clone yourself you'll get someone who looks just like you but who doesn't necessarily act anything like you.

The fascinating concept of duplicate individuals was addressed by Randall Garrett and Isaac Asimov in their ribald ditty, *Clone of My Own*, part of which is:

> Chorus:
> Clone, clone of my own,
> With its Y chromosome changed to X
> And when I'm alone
> With my own little clone
> We'll both think of nothing but sex.
>
> Why should such sex vex
> Or disturb or perplex
> Or induce a disparaging tone?
> After all, don't you see,
> Since we're both of us me,
> When we're making love, I'm alone.

Physical body cloning was possible only after the genetic code was understood. Consciousness is the same for everyone, but the essence, including personality, is different.

This means you can't take full credit for creating a child with a wonderful personality, nor can you take full blame for a rotten kid. But you do have a significant impact on personality development because of three factors:

1. The complete dependence of babies on their parents during early development.
2. The natural desire of a child to "fit in" with their immediate family, partly to benefit from considerations received when others like you.
3. The experiences adult individuals have had that the child hasn't had. Kids have a much better chance for survival if they copy the successful traits of others around them, and those traits will have an impact on their personality development.

A person with a strong personality will seem like a predator; a person with a weak personality will seem like a prey.

Personality Development Factors

A strong or weak personality is developed from the predator/prey tendencies you have, but personality strength is only one aspect of your overall personality. A more important factor is your strength of character. We see a strong personality and strong strength of character in our best leaders. These people recognize values in others, accept appropriate criticism, see the big picture rather than just the special interests, and are not afraid to change their views on previously strongly held positions when convincing arguments are presented. On the other hand, we see a strong personality and a weak strength of character in our worst villains. These people think they're never wrong, surround themselves with people who will agree with them, and accept no arguments against their positions. Given the opportunity, they become fanatics, driven to force others to accept their views and dominance.

A person with a weak personality and strong strength of character will be pushed around by more dominant personality types and hate themselves for being so overpowered. They avoid conflict as much as they can, volunteer for social causes as helpers (not as political activists), are often deeply religious, and try to live quiet, personally satisfying lives. People with weak personalities and weak strengths of character are the dropouts of society. They abuse who they can (bullies fall into this category), lead self-defeating lives where they see themselves as victims, and are generally perceived as losers.

Of course, personality types don't fall only into four neat boxes, but are on a continuous curve. We see all variations, and people can change their personalities under special situations or through a personal wish to change. The pressure of leadership can crack a person's strength, causing them to pursue simpler, less powerful lives. Similarly, amazing heroics are seen under extreme stress, showing inner strength a person never knew they had, and often resulting in permanent personality changes. Occasionally, a weak/weak person recognizes the pain and suffering s/he is causing around them, and somehow pulls himself/herself up to a better life. Your personality and your life can change.

You should keep in mind that your personality is strongly influenced by your physical body, habitat, and culture, and the interactions of these components are complex. A person with a powerful voice, large stature, and muscular physique will be inclined to have a strong personality simply because s/he has more of a physical presence over others. Similarly, a person with a delicate voice, small stature, and weak physique will be inclined to have a submissive personality. Aspects from the habitat and culture add their effects, which can be extremely variable depending on when and how habitat and cultural events happen in the individual's life.

Essence Anomalies

Normally your body contains a single essence. Rarely, however, a person has more than one, resulting in multiple personalities. In cases of abuse, psychologists classify *Multiple Personality Disorder* (MPD) as a psychiatric illness caused by extreme trauma during childhood, and now the person is trying to escape his/her life. When an innocent child is regularly subjected to horrible circumstances, that child may "invent" a different personality in their place so the trauma is done to someone "else". The kid still has a single essence but has fictionalized another personality for escapism. If different kinds of traumas are involved, more than one extra personality may be created. See the work by Richard Baer for more information on extreme cases of MPD. However, in MPD cases that don't have a clear abuse history, multiple essences may be a possibility.

Essence Age

Societies and individuals have historically assigned ages from the physical body birthday. Based on this momentous event, at a particular time on the solar calendar you go to school, are permitted to marry, can vote, and can be elected as one of society's leaders. But chronological age is only one measure, and actually the worst one, of an individual's competence. A better measure is the maturity of your essence, containing your experiences to date.

You probably remember someone who was the smartest kid in the class or perhaps someone who was so academically challenged that they were left behind and were a year older than everyone else. Measuring the intellectual age of children was pioneered by Alfred Binet in the early 1900s and, with collaborators Théodore Simon and Lewis Terman at Stanford University, they measured intellect with an *intelligence quotient*, a.k.a. IQ. The Stanford-Binet Intelligence Scale and others (such as the Wechsler Adult Intelligence Scale) are all commonly called IQ tests. The average IQ is 100, with "genius" defined as above 144, "gifted" as 125-144, and "mentally retarded" as 70 and below. Such buckets tend to make people feel good or bad about themselves but some kinds of intelligence aren't so easy to measure, such as musical talent or artistic creativity.

Some people considered to be profoundly mentally retarded are also *savants* who are incredible wizards at skills often involving memory, language, mathematics, art, or music. Sometimes derisively referred to as *idiot savants*, these individuals have a brain architecture that permits them to visualize the topical goals they seek without performing the calculations or processes taught to the rest of us. The structure of their brains and the way they work are currently unknown. Also, the term "idiot savant" is a terrible way to describe these astounding people with special talents far beyond others who have trained their whole lives in those activities (anyone a bit jealous here, maybe?). Other people have had a serious brain illness (such as epilepsy) or a brain injury that's resulted in similar new capabilities. Again, the reason for the advent of these new capabilities is currently unknown. Examples of capabilities include hearing music once, even a long and complicated classical music piece, and being able to play it perfectly (some of these musicians are blind and have had no musical training); multiplying or dividing numbers to thousands of places in their heads; learning completely any language within a week; remembering everything they've ever read or seen; and being able to draw, in three dimensions, an accurate and detailed landscape of a city after seeing it only once. As you might expect, there's been speculation that the great geniuses in history have had such special brains, which is, of course, impossible to prove.

So, not only do we have another type of age, intellectual age, we also have different ways of expressing it. And keep in mind that intelligence is an aspect of the physical body, not the essence.

Emotional age is the one that should interest you the most. As the exasperated 18th century French philosopher Denis Diderot observed, "All children are essentially criminal." In other words, human children are safe to have around only because they're small and weak. Harry Overstreet said:

> It will mean much to our confused and hostility-ridden world if and when the conviction begins to dawn that the people we call "bad" are people we should call immature. This conviction would bring us to the realization of what needs to be done if our world is to be rescued from its many defeats. The chief job of our culture is, then, to help all people to grow up (p.86)

We all know adults who never grew up and children who seem especially mature for their age. Psychologists have found that emotional immaturity in adults is generally caused by unresolved conflicts from when the adult was a child. Imagine a child, immature in both physical and emotional development, trying to get attention. He soon learns that if he screams loudly enough his parents will give him what he wants to shut him up. Wise parents will teach the child that this is inappropriate behavior and show him other ways to interact with people to achieve his goals. But, if his parents just keep giving in when he screams, he never learns other ways and will carry this screaming into adulthood. The adult screaming takes different forms, including bullying, ignoring logic, and, yes, screaming. Or, perhaps, when the child screamed his parents beat him until he stopped. This helpless child becomes outwardly submissive but inwardly hostile and later becomes a not-helpless adult who's still hostile. Children who never learn to suppress their violence become criminals, but those who do learn to suppress it find other ways of expression. Just look at the front page of a newspaper and you'll see a prominent person who's emotionally immature.

People often see other people's children being brought up badly and become concerned that these brats will be the downfall of humanity. But those judgments may be too harsh. Consider the following quote:

> The children now love luxury. They have bad manners, contempt for authority, they show disrespect to their elders.... They no longer rise when elders enter the room. They contradict their

parents, chatter before company, gobble up dainties at the table, cross their legs, and are tyrants over their teachers.

It's attributed to Socrates.

But bad parenting certainly exists and leaves its legacy on the essence of the children. The emotional age of the parents will dictate much of the emotional maturity of their children.

Your Essence, Your Eternity

Chapter 9 – Your Essence Traits

Your essence is reflected in your personality. This chapter describes each personality trait and suggests guidance to improve it. The operative word here is "improve" and the overall intention is for you to institute *process improvement* on your personality. Process improvement is a well-established commercial approach for creating better products. In your case, your personality is your product. Let's look at the traits that make our personality what it is and what we can do to improve them.

Personality

Personality is the set of individual traits within the essence that determine how we act. In contrast, our behavior comes from the hormones, enzymes, neurotransmitters, and chemicals that are released in our bodies and that help shape our personalities. We should keep this distinction in mind whenever we relate the personality and behavior of an individual with their essence.

Your Essence, Your Eternity

But where, exactly, is personality stored, or is it stored, or what the heck is it? We don't have answers to the physical elements of personality, but Nobel Prize hopefuls are working on it. In the meantime all we can say is that personality exists. It's the same problem scientists have with consciousness, and it's pretty frustrating for them to admit they know so little about something that every living thing has.

Personality describes many facets of the essence and reflects the level of maturity within it. But personality traits are also aspects of the type of living thing we're describing. For example, the personality of a dolphin will be characterized by the kind of living thing (physical body, culture, and habitat) the dolphin is. This personality will, of course, be different from that of a parrot because the physical body, culture, and habitat of the parrot are so different from that of a dolphin. Nearly all personality research has been done with humans, and human personality, specifically yours, is what we'll focus on here.

Chuck Gallozzi suggests nine levels of awareness that describe personalities that he calls a person's *spiritual consciousness*. Those levels are:

Your Essence Traits

- **Level 0:** People at this (low) level care only about themselves. Completely selfish, they are not concerned about the consequences of their actions. They do whatever they want, and if others get hurt, so be it.
- **Level 1:** People at this level are nearly as bad as those in Level 0. But the difference is instead of being indifferent they are thoughtless. It's not that they don't care, but that they don't think about their actions. But they are guilty of abrogating their responsibility to consider the consequences of their actions.
- **Level 2:** At this level people are beginning to experience a spiritual awakening, for after doing something wrong they later regret it. Yet despite their regret, they do nothing to make up for the harm they've done.
- **Level 3:** This level is a giant step forward for after misbehaving, they not only later regret what they have done, but also later apologize or try to make up for their misconduct.
- **Level 4:** At this level, the gaps between misconduct, regret, and apology diminish, for immediately after

doing something wrong the perpetrator regrets it and apologizes or makes restitution.

- **Level 5:** At this level, people are still tempted to say or do something nasty, but as soon as temptation arises, they stop and think. If they make a conscious choice not to hurt others, there's no need to apologize. Of course, they could purposely be nasty, but then there's still no need to apologize!
- **Level 6:** At this high level, there's virtually no temptation, for the people at this level see the good in everyone. They love life and are happy with the world.
- **Level 7:** Although they love life and are happy with the world, people at this level also feel the pain that others are experiencing and do whatever they can to lessen the suffering of others.
- **Level 8:** Lessening suffering becomes the only concern for the people at this level. They lead exemplary lives of service. Dr. Albert Schweitzer and Mother Teresa are examples. People at this level love all, including those at Level 0, for like the 6th century Chinese philosopher Lao Tzu they say, "I find good

people good. And I find bad people good when I am good enough."

People react at different levels to different people and to different situations. If a salesman has just cheated you, you may find it hard to see the good in him. If you win a significant award, you're happier with everyone. If you rush your child to the hospital, you become aware of traumatic situations and can appreciate the pain others feel. Your personality is personalized to events and to people.

Personality vs. Behavior

The personality levels that Gallozzi describes are really characterizations of a person's behavior, not their personality. Remember, personality is part of the essence and is not inherited while behavior is part of the physical body and is inherited. Personality is described by the attributes outlined below while behavior is determined by physical characteristics such as hormone levels, neuronal activity, and chemical balances within the physical body.

If you think behavior comes from your environment (nurture) rather than your heredity (nature), you probably agree with the young toughs in the song *Gee, Officer Krupke* from *West Side Story*:

> Dear kindly Sergeant Krupke,
> You gotta understand,
> It's just our bringin' up-ke
> That gets us out of hand.
> Our mothers all are junkies,
> Our fathers all are drunks.
> Golly Moses, natcherly we're punks!

Evan Ratliff reports an experiment in progress since 1959 at the Institute of Cytology and Genetics in Novosibirsk, Siberia to see if behavior comes from heredity or environment. The originator, Dmitry Belyaev, was wondering what caused some animals to become domesticated to humans while others remained wild. He gathered silver foxes (a species that was never domesticated) and bred them, choosing animals with less aggression towards humans for future breeding. By the 6th generation they were acting as loving towards humans as any pet. Their physical body also changed, developing floppy ears, spotted coats, and curly tails, all characteristics reported by Charles Darwin as normal in domesticated animals. A set of extra-aggressive silver foxes was also bred, and those animals remained especially antagonistic towards humans even when domesticated mothers raised them. So, it looks like the young toughs have to blame heredity rather than environment for their bad behavior.

But in biology, things are hardly ever that simple. Candice Stafford and her colleagues at the University of California report that insects infected with the tomato spotted wilt virus spend more time feeding than insects that are not infected. In other words, the virus changes the behavior of infected insects to better spread the virus among tomato plants. It's not that the virus wants to mess with the insects – it only wants to have more access to the tomato plants it lives on.

The malaria bacterium works the same way. Mosquitoes carry this bacterium and we get infected when they bite us. The natural instinct of a mosquito is to bite and run as quickly as possible to avoid getting swatted. But the malaria bacterium forces the mosquito to stay much longer on their victim, allowing more of the bacteria to be injected. Must also make very nervous mosquitoes.

Viruses and bacteria are masters at changing the behaviors of living things that are part of their life cycle, and don't think it can't happen to you. About 90 percent of the cells in your body aren't your cells – they're bacteria we host to make chemicals we need and can't make on our own. Those bacteria have their own goals that, by default, are also yours.

What about identical twins, individuals with the same genetic makeup? Thomas Bouchard, a psychologist at the University of Minnesota who's evaluated thousands of sets of twins agrees with Wendy Johnson et al. that, "It would appear that, with respect to personality, twins are not systematically different from other people." Geneticist Danielle Reed says, "It's very clear when you look at twins that much of what they share is hardwired. Many things about them are absolutely the same and unalterable. But it's also clear, when you get to know them, that other things about them are different."

From the time we're born until the moment we die there are potentials for conflicts between our personality and our behavior. Conflicts arise when our personality doesn't match our behavior, but keep in mind that the production of chemicals by our bodies depends on circumstances and can vary with age, and our personalities evolve as we acquire more experiences. The interaction between behavior and personality, between the physical body and the essence, is dynamic and complex. Sometimes, though, the effects of this interaction are very noticeable, and we may understand them better by looking at a few examples.

Think about the best person you know, someone who's not only a great individual but who's also happy in his/her life. Put aside your jealousy for a moment and consider why they're like that. The reason, according to our view of the relationships between their physical body and their essence, is because their excellent personality is matched with a body chemistry that enhances it. Everything is working together to make that individual achieve the successes in life that they have.

Now consider the opposite kind of individual. Hopefully you don't personally know anyone like this, but there are individuals who do horrible things and express no remorse for them. You wonder how such people can exist but, as it happens, they have just as stable a life situation as that best person you know. In this case, though, they have an evil personality and a physical body that produces an excessive amount of aggressive chemicals. They, too, are comfortable with their lives and don't want to change them. The balance between physical body and essence occurs whenever behavior and personality are in sync.

Your Essence, Your Eternity

Most people, however, have behaviors and personalities that are often not in sync. Let's say you have a compliant personality but the production of aggressive chemicals in your body has resulted in you cheating another person. You know the name of this conflict – it's called guilt. The further apart your behavior is from your personality, the deeper the guilt. Everyone feels guilty once in awhile, but if this is your normal life then you may have an unhappy disconnect between your physical body and your essence. If your body naturally continues to produce aggressive chemicals then either your personality will be changed to accept this kind of behavior, you'll start taking medication to lower the effect of the aggressive chemicals, or a psychiatrist will earn his/her living from you.

In the opposite situation, where you have an aggressive personality but your body produces an abundance of submissive chemicals, you want to be more assertive but you never are. This can express itself by you feeling angry with yourself for being such a wimp. We're not talking about situations where you're put in real danger (physically, financially, etc.), but rather situations where you're encouraged to take a leading role but refuse to do so. The further apart your behavior is from your personality, the

Your Essence Traits

worse you feel about yourself, possibly resulting in genuine self-loathing. Everyone feels bad about themselves from time to time and you shouldn't worry about it unless it's your normal way of living. In that case, self-loathing can feed on itself and make your life always unhappy. Here again, if your body naturally continues to produce submissive chemicals then either your personality will be changed to allow you to accept yourself, you'll start taking medication to give you more confidence, or you'll seek professional help.

The chemical balance in your body can be changed much easier than changes to your personality. If you think you need to change the way you act, consider making more permanent changes by improving your personality attributes. Want to find out how? There are details about each personality attribute later in this chapter.

An interesting experiment was done in 1961 by Yale psychologist Stanley Milgram on an individual's behavior. He wanted to see if people would really harm others just because they were told to. Through newspaper ads Milgram recruited male participants between ages 20 and 50 of all educational levels for a one-hour session, for which they would be paid $4.50. They were randomly given pieces of paper identifying them as either a "teacher" or a "learner". Actually, all "learners" were confederates of Milgram. Ostensibly, the experiment was to measure the effects of punishment on learning, where the teacher would read the learner a word and give him four choices. If the learner (who was communicating with the teacher from the next room) answered incorrectly, he would be given an electric shock. Shocks started at 45-volts and could be increased to the maximum of 450-volts, at the direction of Milgram. Teachers were given a sample 45-volt shock to see what it felt like. Actually, no learner received any shocks. The learners (confederates) set up a tape recorder integrated with the electro-shock generator, which played pre-recorded sounds for each shock level. After several voltage increases, the learners started screaming with pain and banging on the wall separating them from the teacher, and complaining of a heart condition. At 135-volts, teachers became reluctant to continue the experiment, but were assured by Milgram that they would not be held responsible. After several voltage increases with bangs and screams, learners became silent. If teachers asked to stop, they were given verbal directions in this order:

Your Essence Traits

1. Please continue
2. The experiment requires you to continue, please go on.
3. It is essential that you continue.
4. You have no choice, you must continue.

If the teacher still wanted to stop, the experiment was halted. It was also halted after three 450-volt shocks in succession were administered.

The results were startling. Everyone paused at some point to question the experiment. No teacher stopped the experiment before the 300-volt level; 65% of the teachers administered three 450-volt shocks and, although they were uncomfortable doing so, none left the room to check that the learner was okay. Before you say that all men are closet sadists, a later experiment used female participants, with similar results.

The implications for you are also startling. These results show that the personality (essence) of 65% of the people is not strong enough to overcome the chemical signals their body gives to behave in a way that was very wrong.

As Milgram stated in his description of these experiments, "Stark authority was pitted against the subjects' [participants'] strongest moral imperatives against hurting others, and, with the subjects' [participants'] ears ringing with the screams of the victims, authority won more often than not. The extreme willingness of adults to go to almost any lengths on the command of an authority constitutes the chief finding of the study and the fact most urgently demanding explanation."

Personality Characteristics

So, how do we characterize personality? According to the Diagnostic and Statistical Manual of the American Psychiatric Association, personality traits are "enduring patterns of perceiving, relating to, and thinking about the environment and oneself that are exhibited in a wide range of social and personal contexts." These traits are stable, different among different individuals, and affect behavior. Traits are expressed by personality attributes and, as you might expect, different researchers have developed different lists of attributes. Most of these lists were developed for career evaluation purposes, basically because there's more

Your Essence Traits

money in career evaluation than in personal evaluation. Are you a Type A or Type B personality? Are you a "thinking" or a "feeling" person? Attribute lists are useful to describe how an individual acts, but they don't explain why the individual acts as they do. [Note 1]

Personality attributes develop at different rates and at different times. Fearfulness seems to appear suddenly with a child's ability to experience emotions. Empathy develops gradually over time, hopefully for the life of the individual. Sexual and romantic emotions develop with physical maturity and change as the individual ages.

One weakness of any set of personality attributes is the tendency to categorize individuals based only on the attributes, leading to a one-size-fits-many mindset. However, the essence (described by the personality which, in turn, is defined by its attributes) is strongly affected by (and modified by) the physical body, culture, and habitat. So, evaluating personality becomes a complicated process, changing as aspects external to the individual change.

We've briefly referred to some personality attributes, but now we'll look at the full range of them, what they mean, how your brain uses them, and how they define you as an individual.

Understanding Your Personality

To get some idea of the kind of personality you are, you might take a test based on the Jung and/or Myers-Briggs typologies. When you have this general picture, the following actions will help you understand yourself more fully and recognize the personality strengths you want to keep and the personality weaknesses you want to improve.

1. **Identify your values**. Honesty really helps here, because if you ignore a core value that's a real negative, you'll never improve it.
2. **Bring closure to past events**. The more you can put aside situations that have troubled you in the past, the faster you can move your personality towards your future. Of course, some past events are really ongoing events, but the more of them you have the harder it is to move forward.
3. **Avoid avoidance**. If there's a feared situation that regularly occurs in your life that you keep avoiding, you're limiting the scope your life can have. Avoiding situations is also part of the definition of a phobia, which you certainly want to avoid!

Your Essence Traits

4. **Consider what personality traits you want.** This wish list can be accomplished, but altering your personality by shedding old traits and adding new ones requires dedicated effort. After all, you've been who you are for all this time, and now you want to change. Parts of your body and your mind may object.
5. **Live in the present.** You can't change what happened in the past and you shouldn't fear what might happen in the future. Your personality should help you improve your present so that your future will be the best it can be.
6. **Live your own life.** You know those busybodies who always tell you what to do? Don't be one yourself. Each person develops the personality that will suit their life, and their life is theirs, not yours.
7. **Focus on relationships, not material possessions.** Sure, everyone likes to brag about their accomplishments, whether they're an academic degree, a business award, financial success, sports victories, etc., but isn't it even better when other people brag for you? They'll be far more likely to do that if you're perceived as a person more interested in

them than in yourself. This may be difficult for you, considering how great you really are, but the results are worth it. And, interestingly, the more you focus on other people, the more they'll focus on you!

8. **Recognize the relationships between your physical body, essence, culture, and habitat.** These relationships are described somewhat generally in this chapter, and your life will add the specific details you'll need.

Personality Attributes

The following list of personality attributes includes most of Cattell's 16 Personality Factors and most factors from the Zamora Personality Test [Note 2]. A description of each attribute follows the list. These descriptions do not always conform to those from Cattell or Zamora – they're described here to reflect personality features important to the essence.

The numeric quantities used for evaluation represent the relative values for personalities that exhibit those qualities as the way that person usually acts. For example under ethics, if you're someone who almost always goes along with what

Your Essence Traits

your companions want, you'd be classified as "full self-sacrifice". The important phrase here is "almost always." You'll have times when you exhibit a poor personality attribute, but (hopefully) those times are few and should be ignored when evaluating your whole self. Of course, different people exhibit different levels of the same quality. For example, under emotional stability, someone may be "usually angry", someone else may be "always angry", and yet someone else always seems to have "uncontrolled rage". While these are all levels of "angry", should they get the same value? Of course not. As we learn more about how to judge personality attributes a more refined set of values will be developed. In the meantime, do your best – some understanding is better than none.

You'll see that many of the attributes require personal knowledge of the individual, and that knowledge may change when factors relating to the individual change. Consequently, evaluation of these attributes should only be done for yourself, not for other individuals you may know. No matter what you think, you don't know anyone else well enough. Besides, the great philosophers have said that studying yourself is the key to a better life, and the maxim "know thyself" was considered so important by the ancient

Your Essence, Your Eternity

Greeks that they inscribed in on the Temple of Apollo at Delphi.

According to Aristotle, every action we take is meant to increase our happiness or decrease our pain and suffering. He thought that, above all, humans want to be happy in their lives and that the way to be happy is to learn as much as we can about ourselves. The Chinese philosopher Lao Tzu agreed when he said

> Knowledge of others is intelligence; knowledge of self is wisdom. Mastery of others is strength; mastery of self is power.

The characteristics of an essence can be determined by noting the individual values of all of the attributes. Some people may see this as a contest (Ha, ha – my score is higher than yours, so I have a better personality), but doing that results in an amusing conundrum – the people with the best personalities don't compete to find the best personality! And the very act of engaging in such a competition should lower your score on many of the attributes (warmth, sensitivity, maturity, control attitude, aggressiveness, egocentrism, fairness, and socialization). The goal is to improve your own personality, not to improve others. Other people can use their view of your personality as an example,

Your Essence Traits

either positive or negative. As more individuals improve their personality attributes, the overall essence of the world is improved.

As you read descriptions of the personality attributes, you'll have a natural interest to improve aspects of your own personality. Keep in mind two things: 1) Changing your personality or any part of it isn't easy. Ask anyone who's started exercising to lose weight, started reading to improve their minds, or decided to quit smoking. Don't be one of those people who say, "I know how to quit smoking – I've done it a dozen times." Try to enjoy the process and the experience and look to the end result, not the short-term gain. 2) There are many people ready to help you improve your personality, but some of them are rogues who are more interested in improving financial or control aspects of their own lives than personality aspects of yours. Cults are started this way. Some of these people can be very convincing and you should use extreme caution when a proposed therapy will cost you a lot of money or when someone suggests you must adopt their lifestyle to develop an ideal personality.

You'll probably have the most success with any personality change you consider if that change is initiated by

you rather than by someone else, and if it involves the three basic psychological needs identified by Deci & Ryan:
1. Competence (doing well)
2. Relatedness (connecting with others)
3. Autonomy (acting freely)

When you've decided that initiating a change is something you want to do, appreciate that you've had this old habit for a long time and to permanently change it will take some effort. Brian Seaward suggests four practices for nurturing a healthy spirit, but those practices can be also be used to help change an unhealthy spirit. Those practices are:
1. Centering. Go into a quiet room and put your mind at ease with this proposed change.
2. Emptying. Clear your mind of reasons/justifications you have for the old habit.
3. Grounding. Reflect on the positive aspects this change will bring to your personality.
4. Connecting. Initiate the change and find situations to show the world the new you.

Essence Personality Attributes List

Here's the list of 26 personality attributes that reflect your essence:
- Liveliness
- Abstractedness
- Openness to Change
- Philosophical Attitude

Your Essence Traits

- Perfectionism
- Maturity
- Achievement Attitude
- Task Performance Attitude
- Control Attitude
- Vigilance
- Dependability
- Warmth
- Privateness
- Material Attitude
- Risk Attitude
- Leadership
- Socialization
- Ethics
- Morals
- Sensitivity
- Aggressiveness
- Fairness
- Emotional Stability
- Apprehension
- Tension
- Egocentrism

For consistency with contemporary psychological research, these attributes are organized into the *Big Five Personality Factors* of Openness, Conscientiousness, Extroversion, Agreeableness, and Neuroticism, as follows:

- Openness – imaginative vs. practical people
    - Liveliness
    - Abstractedness

- Openness to Change
- Philosophical Attitude

- Conscientiousness – planned vs. spontaneous behavior
    - Perfectionism
    - Maturity
    - Achievement Attitude
    - Task Performance Attitude
    - Control Attitude
    - Vigilance
    - Dependability

- Extroversion – level of engagement with the external world
    - Warmth
    - Privateness
    - Material Attitude
    - Risk Attitude
    - Leadership
    - Socialization

- Agreeableness – compassion vs. antagonism
    - Ethics
    - Morals
    - Sensitivity
    - Aggressiveness
    - Fairness

- Neuroticism – tendency to experience negative emotions
    - Emotional Stability
    - Apprehension

Your Essence Traits

- Tension
- Egocentrism

You'll notice that some of the attribute descriptions appear to have good/bad end points to the range, such as warmth (very distant to very outgoing); others seem to be bad at the ends but good in the middle, such as privateness (extremely private to extremely open); and others seem to be neither, such as liveliness (cautious to spontaneous). Numeric values are assigned for the ranges but personal bias is always a threat, not to mention the cultural bias of the evaluator. After all, you can't expect everyone to be as fair and dependable as you are. Consequently, until a set of numeric assignments has been accepted by a culturally diverse group of essence researchers, all numeric assessments should be considered as biased. These range values are just descriptions of an individual, not judgments of the individual. And don't forget, you're evaluating yourself, not others.

Because the words used in attribute evaluation are often common words with perceived meanings that go beyond the meaning intended for the attribute, a different font is used to indicate the special meaning to be used. For example, the evaluation criteria for liveliness are (cautious, cautious & spontaneous, and spontaneous). They are shown as (**cautious, cautious & spontaneous, and spontaneous**).

Although we've said these attributes are not intended to be judgmental, the following personality attributes include characteristics useful for the Golden Rule: ethics, morals, maturity, philosophical attitude, control attitude, aggressiveness, vigilance, egocentrism, and fairness. Individuals with positive characteristics in those attributes should be praised and individuals with negative characteristics should be improved.

There are six different attitude personality attributes: achievement, material, philosophical, risk, task performance, and control attitudes. While each has its special characteristics, it's useful to understand the attitude attribute in general.

There are two kinds of attitudes, explicit and implicit. Explicit attitudes are self-reported while implicit attitudes are found through psychological testing. For example, everyone states their support for "equality". However, when given a psychological test that involves equality concepts, distinct biases are found. Cornell University psychology professor Melissa Ferguson checked student's expressed feelings towards "thin" and motivations they had to be thin. While their explicit (reported) attitude didn't match their actions (they ate like hogs), their implicit attitudes predicted the amount of cookies they ate. There's a fair amount of controversy about how the implicit and explicit types differ, even where one starts and the other ends, but it seems that early (even preverbal) experiences and cultural biases affect implicit attitudes the most.

Each Personality Attribute

Openness
Imaginative vs. Practical people

Liveliness

Your natural energy level. Range – cautious to spontaneous.

Courtesy of Marriage Can Wait

Evaluation:

Cautious	= 5
Cautious & Spontaneous	= 10
Spontaneous	= 5

Some reasons for being **cautious** or **spontaneous** are age-dependent, with young children being less cautious because they're excited about life and think they're invulnerable, and old people being less cautious because they figure they've lived a long life and they just don't want to worry about things (sometimes called their *second childhood*). Spontaneity in middle age is often related to a feeling that life is passing you by and you'd better do whatever it is you're thinking about doing before it's too late. This is sometimes called your *middle-age crisis*.

Are you a very cautious person? Then your sense of humor is likely to be more subtle and people who don't appreciate dry wit will consider you rather somber and serious. Cautious people try to find out as much as they can about a subject and, consequently, tend to be specialists in their fields. There's a tendency among specialists to become even more specialized, taking them further from the people who manage their work. The result is a disconnect that always works out badly for the specialist. After all, the people with the money are also the managers.

If you're a very **spontaneous** person then you enjoy social gatherings and being the center of attention. Spontaneous people make friends more easily but also get bored more easily, so they tend to have a variety of interests. Spontaneous people tend not to be specialists because that kind of work is extremely detailed, so they focus on fields for generalists that include most jobs that interact with people. If you manage specialists, though, you'll be much better off if you recognize their value and make sure they know it. Nothing can shorten a manager's career quicker than specialist subordinates out to get him.

Keep in mind that problems can occur in your interaction with the opposite type if you don't understand and appreciate them, and even more so if you've drifted to the extreme of your type. There's an old saying, "Specialists are people who know more and more about less and less until they know everything about nothing, and generalists are people who know less and less about more and more until they know nothing about everything." You should understand that people of the opposite liveliness type have valuable talents you don't have, and you should appreciate their ability to do the things they do. The more comfortable you are working with various personality types, the more successful your personality (and you) will be.

Improving Your Liveliness

There are certainly times to be very cautious and times to be quite spontaneous. But if you're always cautious you'll miss a lot of fun, and if you're always spontaneous you'll get in trouble more than you should because you haven't given enough thought to the situations. The best characteristic is to be both, as appropriate. How do you do that?

You can think of this as a kind of "risk analysis". When a company considers a change that could pose a threat to human health or safety or to the financial security of the company, they proceed very cautiously. You can do the same, in your own way. If your situation could pose a real threat to people, places, or things important to you, use a very cautious approach. For less important situations, more spontaneity is appropriate.

Naturally very cautious people have a hard time being spontaneous because it's in their nature to be careful. You can practice with low-level spontaneity; for example, go off for a weekend without knowing what you'll be doing and without making reservations. Similarly, naturally very spontaneous people go crazy trying to be cautious. You do the opposite – before going off for a weekend, make a list of the things you'll do (and do them) and make reservations before you go. In both cases, build on your new flexibility until you're comfortable with your liveliness abilities.

The very extreme liveliness types should recognize how they're perceived. If you're an extreme specialist you're seen as dull and plodding with a fossilized brain. If you're an extreme generalist you're seen as a fickle, superficial egomaniac. If either of those characterizations is OK with you, you're an egomaniac with a fossilized brain. If you don't open up a bit towards the other liveliness type, your life will be less successful and you'll be less happy.

Abstractedness

The approach you take to solving problems. Range – **imaginative** to **practical**.

Your Essence, Your Eternity

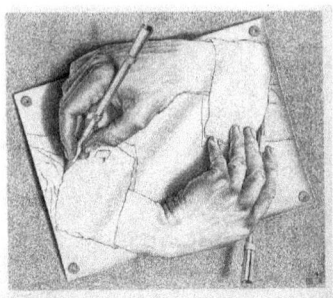

Drawing Hands by M.C. Escher (1948)

Evaluation:

    Imaginative   = 5
    Imaginative & Practical = 10
    Practical   = 5

Are you a big-picture person or a detail-oriented person? If you're big-picture (**imaginative**), you like to propose ideas rather than their implementation, you're good at forming strategic plans, and people sometimes say you're a bit absent-minded. If you're detail-oriented (**practical**), you look to facts to make your decisions, for results rather than outcomes, and ask "how" rather than "why". Imaginative people start companies; practical people are their employees. Imaginative people have to recognize that their ideas come with large risks and large rewards. Practical people have more secure lives, but that security

Your Essence Traits

may keep them from ever becoming rich (or poor!). Some people may have a natural imagination that's gotten them in such trouble that it's suppressed, and those people forcefully inhibit their imagination. Other people may have an innate need for peace and security in their lives but have a wonderful idea that requires they take a substantial risk to complete it. If they drop their idea to retain their security, they should also never think, "If I had only ...". Such thoughts will only make them feel depressed and cowardly. There's nothing wrong with having a quiet, secure life, if that's the life you prefer.

The driving force behind highly innovative people is sometimes hard for the rest of us to understand. There's the story about the failures Thomas Edison had while trying to invent the electric light bulb. A young reporter asked Edison if he felt like a failure and why he didn't just give up. Perplexed, Edison replied, "Young man, why would I feel like a failure? And why would I ever give up? I now know definitively over 9,000 ways that an electric light bulb will not work. Success is almost in my grasp." And, as we know, his persistence eventually led to his invention of a successful light bulb.

Your Essence, Your Eternity

Many parts of life are best handled by imaginative people and many other parts of life are best handled by practical people. You might be able to understand how water flows through the pipes in your house, but if one of those pipes springs a leak, a plumber will have to find and fix it. If you're only big-picture you can easily miss important details in situations; if you're only detail-oriented you tend to see only the trees, not the whole forest.

But there's a time and place for each. If your boss calls you to a strategic planning meeting, put on your imaginative hat. If you're a scientist and you've gotten an unanticipated result in an experiment, you have to think out of the box. If you're looking for a better way to build something, you have to think of a way that no one else has, and that takes imagination. If you're a budding novelist but spend your time on font sizes rather than developing a story line, don't quit your day job.

On the other hand, if you're given a specific task to do, you should be thinking of the most practical ways to do it. If you're a scientist who's writing a paper for publication in a journal, follow their format requirements rather than inventing your own. If you want to become an innovative abstract expressionist artist, you should learn how to paint very detailed scenes first. The finest artists started by learning the fundamentals of their craft. You have to walk before you can run.

Improving Your Abstractedness

As is often the case with personality attributes, a mix of imagination and practicality is the best all-around personality. If you're only the big-picture person you'll pass your bright ideas to others and lose some of the pride of doing the whole job. If you're only the detailed person you'll feel somewhat incomplete because all of your efforts are being done for someone else.

People with **imaginative** abstractedness can add some practicality to their lives by getting involved with the practical implementation of their ideas. This is also quite helpful for your career – the more you know about the specific elements of a project, the more the people doing the implementation will respect you and the better your imaginative ideas will be in the future. Watch out, though – those practical people know you're their boss and may be nervous and/or inhibited. Also, you're used to bossing but must realize the skills of the practical people far exceed your own. Silence is more golden than ever in this situation.

Your Essence, Your Eternity

Similarly, people with **practical** abstractedness can expand their view of the project by trying to learn aspects of the job beyond what they are supposed to do. This is also helpful for your career, because you'll be cross-training yourself to be a more complete employee and be able to better appreciate how your specific part of the project fits in with the whole. Expanding your abstractedness will also make you a more desirable person for other employment. Organizations prefer managers who know what's involved in the tasks they supervise, and they prefer implementers who understand something about the business reasons and goals for the tasks they're implementing.

You can also improve your abstractedness within your preferred problem-solving approach by brainstorming. Big-picture people get together with other big-picture people to toss imaginative ideas around, have them criticized and improved by everyone, and collectively develop the best idea for implementation. Practical people brainstorm with other practical people to figure out how to solve a tough problem. Just by doing this give-and-take with your peers you'll improve your own problem-solving abilities.

Your Essence Traits

Openness to Change

The extent to which you accept new ideas. Range – reactionary to revolutionary.

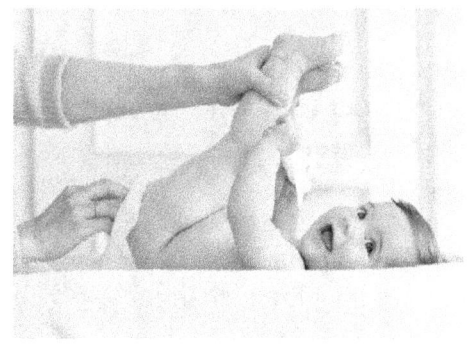

Evaluation:

Reactionary	=	0
Conservative	=	5
Balanced	=	10
Experimental	=	5
Revolutionary	=	0

Interestingly, the Chinese word for "change" literally means "dangerous opportunity". **Reactionary** people focus on the dangerous when a change is proposed while **revolutionary** people focus on the opportunity. The folk singer Bob Dylan expressed the changing atmosphere in the

Your Essence, Your Eternity

United States in the 1960s with his song *The Times They Are A-Changin'*, part of which is:

> Come mothers and fathers
> Throughout the land
> And don't criticize
> What you can't understand
> Your sons and your daughters
> Are beyond your command
> Your old road is
> Rapidly agin'.
> Please get out of the new one
> If you can't lend your hand
> For the times they are a-changin'.

Of course, not all changes are opportunities. Some are just terrible ideas. James Surowiecki found that the best ideas came from groups that satisfied four criteria:

1. **Diversity of opinion**. Each person should have some private information, even if it's just an eccentric interpretation of the known facts.
2. **Independence**. People's opinions are not determined by the opinions of those around them.
3. **Decentralization**. People are able to specialize and draw on local knowledge.
4. **Aggregation**. Some mechanism exists for turning private judgments into a collective decision.

Just because you disagree with changes doesn't mean you're reactionary. If the changes are directly contrary to the Golden Rule you have a moral right to not implement them. If the changes are illegal (some are, you know) you have a legal right to not implement them. But be sure the change is really illegal and not just in-your-opinion illegal. Every year in the United States some people don't pay their income taxes because they say the income tax law is unconstitutional and therefore illegal, and every year these same people go to jail.

People with **reactionary** openness feel threatened by most changes and avoid or even sabotage them. Sometimes reactionary actions arise from self-defense (such as when you feel your job is threatened by the change) or by poor managers who decide to change what you're doing without asking for your participation in the change. A person may also act reactionary to demonstrate their power over others, even others in high positions. The story is that on the day before Eisenhower was to become the U.S. president, outgoing President Harry Truman said, "Poor Ike. Tomorrow he's going to come in and say 'do this, do that' and nothing's going to happen."

A classic example of being reactionary to change occurred in England from 1675 until around 1830 as working situations were being changed from agriculture to industrial style work (the *Industrial Revolution*). Machines were being invented to automate and standardize many tasks, and people started seeing their jobs disappear as these machines were introduced. The *Luddite* movement started in the textile industry in 1811 as a backlash to all new technology, and a significant amount of violence occurred as the anti-technology workers clashed with the police. The Luddite movement took its name from Ned Ludd, although no firm evidence has been found that he was a real person. Since then people who actively oppose new technology are sometimes derisively called Luddites. If you oppose some new technology and find yourself referred to as a Luddite, your opposition will have a much harder time succeeding.

Even if you do succeed in stopping the change, doing it in a reactionary way is a negative personality trait. There are almost always other ways to get changes reexamined and possibly withdrawn. If you really can't stand a change, find a new environment. If acting in a reactionary way is you, you're most comfortable in old, established companies and occupations, such as banks, religious institutions, and governments.

If you have a **revolutionary** view of openness to change you want to change almost everything you see. You dislike

Your Essence Traits

conventions and probably create some turmoil at family reunions, church socials, holiday parties, and other events that include traditional activities. If your life doesn't change fast enough you become frustrated and bored. You change jobs often because you say you want new challenges, but just as much because management knows that changes are expensive in retraining and employee stress, and incredibly expensive if the changes fail. People who frequently suggest changes are high-maintenance individuals and are often the first ones considered for layoffs, even if the reason you were hired was to improve the company. Organizations prefer that no one is indispensable, especially people trying to make others dispensable. You look for new ways to climb that mountain, new places to visit, and new people to aggravate with your need for constant change. You're most comfortable in rapidly changing environments such as the stock market, entrepreneurial companies, and technology research.

But changes happen all the time. If you don't have extreme positions on openness, how do you react? Let's say you're planning a long driving trip to visit various cities. If you make reservations before you reach your destinations, you're **conservative**. If you wait until you've arrived, you're **experimental**. Why is it better to be **balanced** than to be either **conservative** or **experimental** all the time? If you go to a popular site during tourist season, you might not be able to get a room if you don't have a reservation. On the other hand, if you go to a place with plenty of expected empty rooms, making reservations ahead of time will likely cost you more and you may miss a better accommodation. Recognizing how open you should be when you face changes gives you the better result.

More people have **conservative** than **experimental** openness to change. It's easier, it's safer, and you don't have to do anything. People also become more conservative when changes are thrown at them unexpectedly. If your teenage daughter and her boyfriend suddenly announce that they're going to New York City for the weekend, would you tell her to have a good time (experimental) or toss the boyfriend out the front door and lock it behind him (conservative)?

If you're an educator you understand the value of a **balanced** openness to change. Teaching some subjects, such as history and foreign languages, should be conservative because the material is relatively static and methods to teach them have been refined for many years. If you're a teacher with a naturally **experimental** openness to change you'll have a difficult time with such subjects. If you abandon the tried-and-true teaching methods in favor of new but untested progressive teaching, you're more likely to get complaints and unhappy students than praise for your innovations. But other subjects, such as science and technology, change all the time and require teachers who are willing to adapt to those changes. If you're a teacher with a naturally **conservative** openness to change you'll quickly fall behind with the material and student expectations for learning the material. You should either recognize your natural comfort to openness to change and find appropriate subjects to teach or adjust your openness to change to accommodate the subjects you have.

Improving Your Openness to Change

If you're too experimental and want to be more conservative in your openness to change, think twice before you suggest your next change, then don't suggest it. You'll undoubtedly find that the status quo carries on quite nicely, if perhaps not as efficiently as it would with your ideas. Work with the system, as it is, as long as you can. With practice, you should find that your suggested changes are fewer and focused on more significant problems.

If you're too conservative and want to be more experimental in your openness to change, pick the most annoying problem you've encountered and suggest a different way of doing it. Borrowing from Thomas Gaylor's analysis of resistance to organizational change, the following four factors are suggested to improve your openness to change:

- **Participate in the changes**

    Change represents a threat to your personal status because you've finally become familiar with the way things are and now they're changing. Changes are much worse, of course, if you find out about them after the change decision has been made. To minimize the surprise factor, try to become involved in the change-making process. This will allow you to have a greater understanding of the need for the change and give you a partial feeling of ownership in it. The people involved should appreciate your honest interest in making the best change possible, but they won't appreciate discovering that a change saboteur has infected the group. Work with them or you won't work at all.

- **Trust in the people around you**

    Face it – if you don't trust the people making changes, you won't trust the changes they make. This lack of trust could be because the people have been insincere, unreliable, or just plain dishonest, but it could also be a lack of understanding on your part. If you're stuck with these people and could be stuck with their potential changes, the best approach is to become involved yourself. Try your best to wipe your mind clean of negative thoughts about those people and ask to participate with them to develop the changes. Smile when you do.

- **Communicate**

    Good communications are vital to successfully implementing changes, and those communications start with you. If you're confused about a change, ask questions. If you're upset about a change, talk about it. If you don't like what someone is saying about a change, get involved in the discussion. How you express your views about changes determines how much your views will be considered. Some people are naturally good communicators, and those people will be heard – good for them. If you have communication difficulties, try to find some way to express yourself. Your proper openness to change is completely dependent on how much you understand the change. Also, if you disagree with a change, there are always communication avenues for you to pursue to have it reexamined.

- o **Obtain good information**

    If you get your information from unreliable sources, the conclusions you make from it will also be unreliable. Technology developers use the term "garbage in, garbage out" to describe what happens when you use bad data. If you're tricked by a dishonest person who gives you incorrect information to make you believe something, all you can do is remember that the person is not to be trusted again. But you basically know which sources are good and which are bad, and it's up to you to choose those that help you understand changes. If you purposely select information that fits your interests in making or not making the change while ignoring contrary information, you're being unscientific and your conclusions will be seen as self-serving.

Philosophical Attitude

The extent of your willingness to change your attitude. Range – **fossilized** to **rambling**.

Evaluation:

    Fossilized = 0
    Defensive = 5
    Flexible = 10
    Rambling = 0

Do you think that we're animals that have evolved as an accident of nature, or do you think a Divine Creator put us on Earth? If the position with which you disagree presents evidence, are you willing to consider it? This is the basis of your philosophical attitude – how willing are you to change your opinion. Philosophical attitude refers only to opinions, not facts. If someone says the sun rises in the east, that's a fact, not an opinion. If someone says that children are never too young to be spanked, that's an opinion, not a fact.

This is a good time to get dictionary meanings for "fact" and "opinion":

Fact: something that actually exists or has actually occurred; something known by observation or experience to be true or real.

Opinion: a conclusion or judgment held with confidence, but falling short of positive knowledge.

If you believe something to be a fact and it's challenged, ask to see the evidence. There are limits, of course. The Flat Earth Society still exists and will show you their evidence that the Earth is flat, irrespective of what you've learned in school. Much of science involves challenging established facts, but it's done with experimentation using the scientific method. You have to decide what's pure nonsense and what deserves your consideration.

A **fossilized** philosophical attitude means the person forms an opinion that never changes, no matter what. It's

Your Essence Traits

easy to have a discussion with a fossilized person because you're never confused about their position. You might be frustrated at how completely your perfect logic is discarded, but you have to admire someone who knows with such certainty that they're right. Fossilized people are at the extremes of any argument, both for and against. If you don't want your stomach problems to get worse, avoid talking with them.

A tree becomes fossilized by having its organic matter slowly replaced by stone. A similar process occurs with industry and government leaders who surround themselves with "yes men". After all, if everyone around you agrees with you, you must be right, right? James Surowiecki points out the fallacy in this thinking throughout his book and says, "One of the quickest ways to make people's judgments systematically biased is to make them dependent on each other for information." (p. 41)

A **rambling** philosophical attitude is no better. If you're a rambling person you change positions often, usually to the exasperation of the people you're talking with. You tend to agree with whoever spoke last. This was amusingly illustrated in the musical play *Fiddler On The Roof*:

Your Essence, Your Eternity

> Townsperson: Why should I break my head about the outside world? Let the outside world break its own head....
> Tevye: He is right...
> Perchik: Nonsense. You can't close your eyes to what's happening in the world.
> Tevye: He's right.
> Rabbi's pupil: He's right, and he's right. They can't both be right!
> Tevye: (Pause). You know, you are also right.

If you ramble, people will avoid you. You certainly don't realize people think you're an indecisive rambler or you'd do something about it. Think about what you say because if you don't know you're rambling then ...

A **defensive** philosophical attitude is a real improvement, although by adopting it you advertise your skepticism about the arguments presented without real evidence they're wrong. Perhaps you've developed a defensive attitude by being hurt too much and too often by being lied to. Or maybe you just like to disagree with people. On the plus side, people have made paid careers of being skeptics, but you'd better have good arguments to support your skepticism or your career won't last long. The environment for your skeptical attitude matters, too – people who have disagreed with absolute rulers have sometimes wished they hadn't.

The best kind of philosophical attitude is one where you're flexible to new ideas. If you have immediately rejected that last sentence, go back and reread about **fossilized** philosophical attitudes.

Being flexible doesn't mean you accept a person's arguments, just that you don't reject them out of hand. Some people are born with an open mind to various opinions, but most of us aren't and to have a **flexible** philosophical attitude may require a significant amount of effort. For example, if someone says that children are never too young to be spanked and you say they're an idiot, you're not being flexible. But if you cite, for example, a study published in the September 2009 issue of *Child Development* that says spanking children less than 3 years old is bad for them and then say the person is an idiot, you're being flexible. Watch out, though, because no one likes being called an idiot and he may come back with the Bible verse Proverbs 13:24, "He who spares his rod (of discipline) hates his son, but he who loves him diligently disciplines and punishes him early", and suddenly you find yourself in a messy argument. Better to state your opinion without extra adjectives, listen to other opinions pro and con, and then make your own judgment.

Improving Your Philosophical Attitude

If you have a **fossilized** philosophical attitude the most important thing is to recognize that you have it and accept that it's not a good way to interact with people. The recognition part may come from friends who chide you for being so closed-minded. People tend not to say such things without reason, so should someone say that, think about it. If more than one person says it, think about it really hard. At this stage you should try to change your thinking to being skeptical rather than being completely sure. Ask for evidence for the opinion you disagree with and then see if you can live with that evidence. If the evidence is too weak, you've made a judgment based on facts rather than on attitude and, at worst, you're being defensive.

A **rambling** philosophical attitude is easy to see because you know how hard it is for you to make decisions. To improve yourself, consider each of the arguments you hear separately, then pick the one that seems best to you and stick with it. Of course, if new information is found that adds significant weight to another position you can change your mind, but make sure your change is for a good reason and really stick with your new position. You should also start to feel better about yourself as a person, as someone who isn't that easily manipulated. Be careful, though, because it's relatively easy for a former rambling person to become a fossilized person.

If you have a **defensive** philosophical attitude and want to be less critical, following the *tit-for-tat* strategy mentioned in Chapter 7 may help. In this case, you'd accept the person's arguments the first time but if you felt betrayed you'd demand proof for his/her next argument. If someone thinks you're easy to fool not only will you often wind up in a poor position but you'll also lose the respect of your peers. Do to others ...

Conscientiousness
Planned vs. Spontaneous Behavior

Perfectionism
The extent of your acceptance of disorder. Range – **careless** to **compulsive**.

Your Essence, Your Eternity

Evaluation:

Careless	=	0
Spontaneous	=	8
Orderly	=	10
Systematic	=	8
Compulsive	=	0

It's easy to generalize that extremely orderly people are high achievers while sloppy people are low achievers. Certainly the extremely orderly people think so, but it's not true. The perfectionism trait reflects your clutter comfort level, nothing else. The following exchange illustrates the natural conflict between orderly and disorderly people.

A manager who cleans his desk at the end of each day comes into a worker's office where litter is everywhere:

> Manager: A cluttered desk is the sign of a cluttered mind.
> Worker: What's an empty desk the sign of?

Your Essence Traits

Your comfort level may also reflect a dramatic, if temporary, change in your life. For example, if you're normally a very orderly person but a loved one living far away is now very ill, you may be so distracted with concern that your own life patterns become secondary and the order within your life becomes unimportant. In the opposite way, if events in your life have become confused and uncontrollable, even a disorderly person will try to find some things that s/he can control and will make them as orderly as possible. When you're trying to evaluate your perfectionism, you have to take your life's current situation into account.

If you're evaluated as **careless** you're disorganized and regularly leave things lying around. You clean only when forced to by others and take your time doing it. In sporting and recreational activities, you attempt difficult things, without hesitation, that no one else will try. You've had a traffic accident and have broken at least one bone. If you're living with someone else, that other person should also have a **careless** approach to perfectionism or there will be frequent strife in the relationship. If you're normally late when meeting people, can't finish your tasks on time and always have some excuse, you probably have a **careless** attitude towards perfectionism. Interestingly, these traits also apply to some of the greatest geniuses in history, but you get a zero evaluation nonetheless. That's because (as you recall) intelligence is a physical body trait but personality is part of your essence. Being careless is a bad personality trait.

If you're a **compulsive** person, on the other end of perfectionism, you're so organized that you're inflexible to the point of paralysis. "A place for everything and everything in its place." You follow rules as precisely as you can, read all the instructions and do all the steps in order, and observe the letter of the law. If you also adhere to the spirit of the law, it's a happy but unintended consequence. If you're compulsive, the monetary tips you leave for service people are calculated to the penny. When you travel you have, of course, made a schedule of activities that must be followed. Adhering to this schedule occasionally results in your rushing to keep on time (not everyone you interact with will be as compulsive as you) and missing serendipitous events you may encounter. Your rigidity tends to keep others away from you. Compulsive

Your Essence Traits

people make good soldiers (but not good officers) and are most comfortable in very structured environments.

Spontaneous people still have a flexible attitude towards deadlines but they don't act so cavalier about it. If you have a **spontaneous** attitude towards perfectionism you "work to live" rather than "live to work." You won't hesitate to try new ideas even if they threaten the schedule or even the success of the activity. Consequently, you usually require closer supervision if a timed task is involved. In competitive events you'll try that tricky shot, do that extra flip, or paddle through that dangerous white water. You enjoy extreme sports but don't try foolish things. Well, maybe a little foolish, but not idiotic things. You're most efficient when your environment is unstructured and would be good as an artist, inventor, or entrepreneur. The risks you take sometimes result in advances we all enjoy. Being spontaneous adds excitement and a fair amount of enjoyment to your life, even when things go wrong.

Systematic people may not be the entrepreneurs, but they're the people an entrepreneur relies on to get the job done. You're most efficient when your environment is structured and you "live to work" rather than "work to live." You take deadlines seriously and follow the process, but recognize when the process must be changed to accommodate a special situation. You've probably charted a career path for yourself. You prefer organized activities and have a hard time handling teenagers. You take violin lessons rather than fiddling around by yourself. It's sometimes easy for you to fall into compulsion and wind up missing social and recreational aspects of life.

People with an **orderly** attitude towards perfectionism combine the best features of the spontaneity and systematic traits. If you're orderly you balance creativeness with organization to be able to propose something new and see it to completion. You also make everyone around you feel pleased with how it was done. You work hard but know when to relax.

Here are some traits of perfectionists. Decide for yourself how you act for each of them:

1. High goals are set that must be met. No excuses.
2. Small imperfections seem more important than the overall accomplishment.
3. Fear of failure is paramount.
4. Criticism is not accepted.

Improving Your Perfectionism

If you're **careless** you need to put some structure in your life before the Board of Health condemns your home.

Your Essence Traits

Also, if you're not one of the great geniuses in history you probably have trouble keeping a job. Since you have such trouble with organizing and organizations you should begin simply by cleaning one room in your home and start brushing your teeth regularly. Make a list of everyone you know with their addresses and contact information. After you've successfully completed those projects, expand them to more extensive cleaning and more directed organization in your life. Invite a friend who's not careless to assess how you're doing. Other aspects of your life should be easier when you've learned to do basic organization and hygiene.

If you're **compulsive** you need to loosen up. Excessive compulsion may be an indication of low self-esteem – you're trying to make up for some perceived inadequacy by being perfect in other things. There's certainly nothing wrong with high goals, but try to understand that perfection is usually impossible. Choose something you're trying to achieve and ask a friend you respect to suggest a high goal for it. Work towards that goal, complimenting yourself as you make progress. The key phrase here is "work towards that goal". Compulsive people feel compelled to achieve the goal, but now you're just working towards it. If the goal is never met but you've given your best effort to do it, you're a success.

It was already mentioned that an **orderly** attitude towards perfectionism combines the best attributes of the spontaneous and systematic attitudes. So, if you're too spontaneous put aside your more dramatic suggestions in favor of more structured approaches. If you're too systematic look closely at the processes you have and suggest improvements to only the ones that need it the most. Since being spontaneous and being systematic are both valuable aspects of **orderly** perfectionism, a spontaneous person should continue being spontaneous while adding more standardization, and a systematic person should continue being systematic while adding more spontaneity. You shouldn't completely change who you are, especially to improve your perfectionism.

Maturity

How much you consider consequences in making decisions. Range – **immature** to **wise**.

Your Essence Traits

Evaluation:

    Immature =   -2
    Infantile =   0
    Responsive =   5
    Thoughtful =   8
    Wise =   10

There are several kinds of maturity when referring to living things. There's *physical maturity*, when the individual has reached its reproductive age; there's *mental maturity*, when the individual is able to understand the concepts expected of an adult; and there's *emotional maturity*, when the individual behaves in an adult manner. Physical and mental maturities are aspects of the physical body; emotional maturity is an aspect of the essence and it's the one we're looking at here.

Physical development in the brain is closely linked to the emotional maturity of the individual. A child is not able to act as emotionally mature as an adult because his/her brain hasn't physically matured to that stage. So, when you see a child throwing a tantrum at a toy store, you know that the screaming is part of the child's undeveloped brain (and part of the underdeveloped child-rearing capabilities of its parents). After adolescence, though, the brain is physically fully formed and all capabilities of adult emotional maturity are available to individuals who don't have damaged brains. When we evaluate the maturity of an individual, then, we have to consider the physical age of that person.

Infantile individuals throw tantrums. Tantrums can take different forms – the immaturity of the small child screaming on the toy store floor is no different than that of an athlete who threatens an official for a perceived bad call, or that of a politician who won't stop talking until he gets his way. If this sounds like you, perhaps the excitement of the moment generates a chemical imbalance that causes a temporary loss of maturity. If you later recognize and recant your poor behavior, that immature event can be forgotten. If it continues to happen and you continue to apologize, though, you need some more permanent maturity improvement.

A person with an **immature** attitude towards maturity takes a narrow, selfish approach to decision making, irrespective of its effects on others. They're not necessarily evil but they're thoughtless to everyone except themselves. The biggest concern, and the reason they earn a negative evaluation score, is that they can be very dangerous to innocent people around them while honestly believing they're doing the right thing. These are the people who "are only following orders" and who destroy another person's livelihood because of suspicions about them, without considering facts. They think they're never wrong, get advice only from people who agree with them, and if they say something it's true just because they said it. They can be completely tricked if a crook has a convincing pitch, but if something goes wrong the crook could be in serious trouble. Immature people are blindly loyal to a political party and always vote for its candidates without caring what those candidates actually stand for. If the leaders they follow encourage violence, they'll follow with righteous belief in whatever they're told to do. They'll condemn another person as immoral for having a sexual affair, but if they have one it was only a "temporary weakness of the flesh". Be careful if you associate with an immature person because they'll remember things about you that will benefit them in the future.

Your Essence, Your Eternity

Responsive people make decisions mainly towards their own needs, but when people they know have opinions on the issues they listen and may change their own decisions accordingly. Does this sound like you? You're quite responsive to marketing pitches and celebrity opinions. You don't pay as much attention to the opinions of recognized experts because all sides of an issue have their own experts and you prefer relatively straightforward analyses. If you're under financial stress or just plain greedy, you're a prime candidate for scams. The saying "if something seems too good to be true, it usually is" applies here. Both defense attorneys and prosecutors try to choose responsive people for juries because they think they'll be able to sway your opinion better than their opponent can. Political candidates focus their campaigns towards responsive people, and you're the people who constitute the bulk of the undecided voters.

Thoughtful people have the same attitudes and attributes as responsive people but try to expand their understanding of issues by listening to more opinions from various sources before making their decisions. Celebrity opinions are included, as are the opinions of the various domain experts. Does this sound like you? While personal needs are first in your mind when making decisions, consideration is given to the consequences of that decision on people you don't know and on the world in general. You remember and try to acknowledge events special to other people (birthdays, holidays, etc.). You help little old ladies across streets. If you need to borrow something from a neighbor, go to a thoughtful person first.

Marketing to thoughtful people has to include enough substance to convince them that there's a real difference in the product. An example is the successful approach IBM used during 1964-79 for its mainframe business computers. In this relatively early period for commercial computers they built the IBM/360 for sale to government, industry, and academia. At this time the industry decision-makers who bought such expensive equipment weren't experts in technology and IBM followed a brilliant marketing strategy by adding hundreds of lights flashing different colors on the front panel. The lights indicated activities going on inside the computer, but they flashed so quickly that the front panel looked something like a lighted Christmas tree. Potential buyers were very impressed ("Look how hard this computer is working!") and the 360 sold quickly. To be fair to IBM, the 360 was an excellent computer that did the work it was intended to do, but the marketing approach was instrumental in making the machine so popular. Later, the buyers had become more knowledgeable and realized that such lights were more of a gimmick than a tool, so in the next version (IBM/370), all of them were eliminated. Instead, the buyers saw a flat black front panel with two buttons – Start and Stop.

People with **wise** maturity consider the consequences of their decisions on the world first with their personal needs as secondary. Does this sound like you? You'd rather give than receive, volunteer for relief services, and are generous with charities. You spend a lot of time analyzing issues to not be tied to one position until you have all the facts. You understand, as James Surowiecki found, that "under the right circumstances, groups are remarkably intelligent, and are often smarter than the smartest people in them." (p. xiii) Consequently, you look to the opinions of diverse groups to help form your opinions. While you work as hard as anyone on your job, you do it for reasons beyond the monetary compensation. A rich person can appear to have a wise maturity level by giving away money and volunteering their time, but only they know if their decisions are as unselfish as they seem. The world would be a better place if all its leaders had a wise level of maturity.

Here are some important behaviors associated with maturity:

- Ability to control your anger and settle differences without violence
- Respecting other people's opinions
- Becoming less self-centered
- Having patience while you build a better future
- Accepting defeat
- Ability to graciously and sincerely admit you were wrong
- Accepting the consequences of your decisions
- Assisting others in their times of crisis

Improving Your Maturity

Here are some suggestions for improving your maturity:

- **Be realistic**. Daydreaming is fine but include concentration, attention to detail, awareness of the real possibilities, and dispassionate analysis of information.
- **Challenge your assumptions**. Many of your beliefs are from your childhood, from friends and family, and from your previous encounters. New information may have changed your basic beliefs and this

information should be seriously considered in your decisions.

- **Be inquisitive**. Find diverse people you respect and consider their opinions about an issue before making your own decision.
- **Listen to others**. People you respect can also help you improve your maturity by pointing out mental blockages you have. Ask for their help, be grateful to get it, and use their thoughts to consider changes in the way you make decisions.
- **Understand your emotions**. Nobody expects you to completely control your emotions, but you should be aware of elements in them that make other people uncomfortable. Make a serious effort to discover what triggers those elements and ways to cool down your reactions.

Achievement Attitude

The extent of your motivation to reach goals. Range – **reluctant** to **obsessive**.

Your Essence Traits

Evaluation:

    Reluctant =   0
    Indifferent =   2
    Motivated =   8
    Focused =   8
    Obsessive =   0

Everyone has goals they want to achieve, and the way they do it indicates more than just their achievement attitude – it also reflects their perfectionism, risk attitude, and task performance attitude. Desired achievements also change over time, with career goals prominent when you're young and lifestyle achievements when you're older. Every achievement comes with a cost and your willingness to accept that cost becomes the lifestyle you set for yourself.

Your Essence, Your Eternity

Do you want that college degree? You're going to have to sacrifice money and social time (at least) to get it. If you decide you'd rather go out with your friends than hit the books every night, you're setting a different direction for yourself. If you've made the social choice rather than the academic choice after a thoughtful analysis of the kind of person you are and want to be, then you have a sound achievement attitude towards being a party animal. You understand you're probably sacrificing career options and higher pay, but you've decided that the social sacrifices are too great. However, if you've made that choice because you know studying takes a lot of time and you just don't want to work that hard, you're a slacker with a poor achievement attitude.

Some people have a high achievement attitude for things that shouldn't be achieved. A museum has a work of art that you'd love to have, so you form a plan to steal it. You want to get a high grade on a school test and think your best chance is by cheating. You want to make more profit on the products you build, so you secretly start using cheaper, inferior materials instead of the materials you've used in the past. While your achievement attitude may be

Your Essence Traits

high, your ethical and moral levels will suffer and you'll be seen as a greedy, selfish crook.

If you have a **reluctant** achievement attitude you're satisfied with achieving as little as possible. You might not be just a bum – you might have physical, cognitive, or emotional issues such as a learning disability, an attention deficit, emotional disturbances, psychological disorders, or other health impairments. After these physical ailments are treated, your achievement attitude can be addressed. Of course, if you don't have a physical problem that's causing the low achievement attitude, you may be just a bum.

Indifferent people don't care if they do well or not. If this sounds like you, you're an underachiever who intentionally doesn't try very hard at your activities. For example, you may play a game such as chess that requires a lot of concentration but spend your time talking to friends or daydreaming, and then wonder why you keep losing. You also annoy players who are focused because you take away the competitive fun of the game (unless you're playing for money). You're used to losing and are a good sport when you do. However, other good sports who have tried their best but who are just not as talented as their competition are not indifferent – they're just good sports. Indifference could be a symptom of low self-motivation or self-confidence, in which case those issues should be addressed first.

Motivated people try their best to achieve their goals, but those achievements aren't central to their lives. You set sub-goals on the path to achieving your primary goal and, if achieving the main goal becomes too difficult, those sub-goals are satisfactory end points. Does this sound like you? This is not to say that you're a quitter, rather that you set a series of smaller goals that can be accomplished within the time and resources you feel are reasonable. You're in the group that makes the best managers because you understand what can be done and your own motivation stimulates your colleagues to achieve those goals. You always have two or three things going on at the same time and have a buoyant, happy lifestyle. You organize community activities, are active in the public schools, spend time playing with the kids, and have dinner parties every other week in addition to working full time. Having a motivated achievement attitude is fun and stimulating, but some quiet time is also needed to recharge your batteries. Take a nap.

You may have noticed that the same evaluation score is given to both **motivated** and **focused** achievement attitudes. This is because motivated people sometimes find

Your Essence Traits

a project or activity that they especially like and focus most of their energy on it. While this is generally a limited time activity, for that limited time the motivated individual becomes a focused individual. People who are naturally focused in their achievement attitude prefer to be responsible for completing tasks but don't feel comfortable in managing them. Focused people have just as high an achievement attitude as motivated people but direct their energies towards fewer projects. You need to be focused to achieve some things, such as learning a part as an actor, doing dissertation research, or inventing something new. Focused people are the technical experts in activities and, if paired with a motivated manager, form an excellent team.

If you take your focus to an extreme (and sometimes unhealthy) level, you have an **obsessive** achievement attitude. There's no such thing as achieving perfection to you because you're always trying to do the activity better. While no one faults self-improvement, when it takes over your life to the extent that you lose touch with others (and even with reality) your self-improvement becomes self-destruction. A world-renowned chess player may make the news because of his or her obsessive behavior, but other people who have taken the identity of a character in a game or who are never away from their work are just as obsessed. There are certainly times when focused people become completely involved in their project, but if that focus lasts too long then you've become obsessed. You often don't realize what's happening and it's up to your friends to help.

Improving Your Achievement Attitude

Achievement attitude has a direct relationship to your view of how you're doing in your career. During World War II sociologists headed by Samuel Stouffer from the University of Chicago looked at military police (who were rarely promoted) and airmen (who were frequently promoted). To their surprise, the police were happier in their jobs than the airmen, possibly because the policemen compared themselves with colleagues who hadn't been promoted while the airmen compared themselves with colleagues who had been promoted. Other studies found similar results in salary comparisons – not the actual salaries, just the comparisons with co-workers. Jealously impairs achievement attitude – just do the best job you can and enjoy the satisfaction you get from doing good work. If you get promoted because of it, congratulations!

Your Essence Traits

Since neither **reluctant** nor **obsessive** people will get to this point in this book (reluctant people are too lazy and obsessive people won't take the time), it's up to their friends to help improve their achievement attitudes.

People with a **reluctant** achievement attitude should be assessed by a professional to see if they have any of the physical impairments that could cause their attitude. If the assessment turns out negative (i.e., the person is a bum), the best way to help them is to find something they want to be good at. Even the most seemingly useless person is good at something, and the goal is to find it and encourage the person to get even better. This may not always work, but if a person retains a reluctant attitude to achieve they will be a burden on everyone around them and on society as a whole.

Helping an obsessive person involves getting them away from their obsession. Since they're devoting their lives to this one activity, getting them away from it to do something else sounds very difficult, but it's not. The key is pointing out how doing another activity will actually help them when they do their obsession. For example, if someone is obsessed with the Internet and never leaves their computer, point out studies that show a moderate amount of consistent exercise stimulates the mind and leads to a longer life. Better still, look up those studies on the Internet. Then offer to go on a nice walk in the park with your obsessed friend, for just a few minutes. Once you're able to get them to do something else, it's easier to expand those activities to others in which they show interest. Keep in mind that the idea is not to change their personality, it's just to broaden the perspectives in their life. If the person eventually develops a focused achievement attitude, you're a success.

If you're indifferent or motivated or focused and want to change a bit, a new achievement attitude can be developed once you understand how capable you are at successfully performing a task. The term for this is *self-efficacy*. Achievement improvement depends to a large extent on your willingness to want to improve, and to do that you have to know where you're starting. Then you should develop a way of monitoring yourself and institute performance changes that improve your achievement results. Make sure your friends congratulate you on your improvements and have them encourage you to achieve even more. Getting your friends to do that will be another achievement.

Your Essence Traits

Task Performance Attitude

Your work ethic. Range – careless to obsessive.

Evaluation:

Careless	= 0
Casual	= 3
Methodical	= 10
Determined	= 5
Obsessive	= 0

The two criteria important to task performance are ability and attitude. Both must be present – if you know the job but don't want to do it, you'll fail; if you want to do the job but don't know how to do it, you'll fail. Since we're looking at personality attributes, we'll only discuss your attitude and assume you know what you're doing.

Your Essence, Your Eternity

How do you know if you have a good work ethic? You can usually tell by the things you do when you're not working. For example, if your favorite personal activity is watching TV, uh-oh. However, if it's a participation sport (baseball, tennis, swimming, etc.) you show a self-improvement initiative that's likely to continue in tasks you do. Besides, you're getting healthy exercise and have an outlet for the frustrations you acquire while doing tasks.

To get a job (and to keep that job), you should have shown a positive work attitude during the recruiting process. Vonne Meussling points out that task performance attitude is part of the way your workplace associates view you, including how you act during training and in your overall morale on the job. Your task performance attitude is also shown in your job productivity and output, work performance, absenteeism, and your organizational commitment.

What kind of a task performance attitude do you have? If it's **careless**, you probably care less about the task and you're anxious to do something else. We all have tasks in our lives that we'd rather not do, but if you notice that you dislike most of the things you do when you're working, you're wasting a significant part of your life and not helping

Your Essence Traits

people who rely on you to get your tasks done, and you'd better change your attitude. However, if everyone involved in the task is careless, the task is probably so onerous that it was invented as a place to get rid of unwanted employees. But if only you are careless either everyone else is semi-catatonic (unlikely) or it's just your bad attitude. Change it.

Having an **obsessive** task performance attitude isn't any better. Then you're a workaholic who drives everyone crazy with your insistence on perfection. You also like to work alone because (you think) no one else can do the job as well as you can. This level of intensity is dangerous because you don't do anything else, including eating and sleeping, until the task is done. A sign in a computer software company illustrates this attitude, "There comes a time in the life of every project when it becomes necessary to shoot the programmer and start production."

Being an obsessive individual is bad enough, but if you're an obsessive individual who's supervising others, it's worse. You tend to micromanage, making your coworkers feel inadequate and tense, which sometimes results in unfortunate remarks on both sides. The result, of course, is an overall lower performance in accomplishing the task. There are cases when an obsessive person is also an inspiration, the heart and soul of the organization. An example was Steve Jobs of Apple Computer, who was so demanding and so inspirational that it was said his employees loved him and hated him at the same time. This is not something you should try yourself – usually people just wind up hating you.

If you have a **casual** task performance attitude you're friendly and easy-going, making you an excellent person to work with, but you apply too much of those desirable traits to doing your tasks. If the task has a tight schedule (for example, you can't put up the walls of the house until the electrician has installed the wiring) or if extreme focus is required for the task ("nurse, sponge stat"), it'd be better to hold your easy-going approach until later. On the other hand, your friendliness makes you nice to have around. There's the story of a company that needed to reduce their staff and upper management directed the supervisor to suggest someone to be laid off. The supervisor noticed a woman on a team who didn't seem to be doing anything and thought she'd be a good candidate for reduction. In looking at her employment history, the supervisor found that she never seemed to do anything on any team to which she was assigned, but all of those teams completed their tasks within the schedule and under budget. In fact, only teams with her on them completed their assignments efficiently. It turned out she was a "catalyst", a person who didn't contribute directly herself but by having her on the team everyone else performed at their peak. She was certainly a special person, and one who illustrates that someone with a casual task performance attitude may be contributing more than it seems.

If you have a **determined** task performance attitude you're a no-nonsense type determined to get the task done. When you play you're just as determined to play well, and when you're having fun, by golly, you're determined to have fun! Sometimes over the top, you're the one to select if something needs to be done and the one to invite as the life of the party. Often more involved with yourself than with others, it's rather easy for personality clashes to develop. Fortunately, though, if someone takes you aside and explains the difficulty, you'll be determined to make everyone happy again. If another determined person is in your activity and you disagree, you may have task performance difficulties. You're certainly not high maintenance, but you're also not casual and will require some "supervising" attention.

Your Essence, Your Eternity

A **methodical** task performance attitude is the most desirable because it combines the best aspects of the casual and determined types. A step-by-step approach with a proper amount of advance planning is the best way to get tasks done in the most efficient way while keeping the people involved as happy as possible. The most successful methodical people know that advance planning is very task-specific, sometimes with only the broadest understanding of major aspects of the task in the early stages. As with every other activity, the more experience you have with performing tasks the better you are at it, with more knowledge of when and how to be methodical.

Americans sometimes think Europeans are too casual in their task performance attitude, while Europeans sometimes wonder why Americans rush around as much as they do. There is a difference between activity and accomplishment, recognized by methodical people who follow the adage, "If you don't have time to do it right, how will you ever find time to do it over?"

Your Essence Traits

Improving Your Task Performance Attitude

Sometimes improving your task performance attitude only requires a little guidance. Four main types of guidance are:

- **Additional exposures** to the right way to do the task. Find someone who knows the task very well to show you how they do it.
- **Positive reinforcement** to reward desirable results. Ask people to provide feedback on your performance, but ask them to follow the Golden Rule (Do to others …). You don't need a lot of negative bashing.
- **Persuasion**, similar to methods advertisers use, to convince you to do something a certain way. Persuasion only works when the person doing the persuading is someone you respect.
- **Positive views of the task** to make it seem more desirable. Sometimes you can develop a positive view by thinking of the alternative, namely not having the job (and paycheck) at all.

If that guidance isn't enough to improve your task performance attitude to the level you want, here are some suggestions.

If you generally have a **careless** task performance attitude, you have to find something that stimulates your interest and parley that towards improving your attitude on your tasks. By doing that, the people around you will see you as a new positive person and you'll get more attention, better tasks, and a generally happier life. If you don't find something to change your poor attitude you'll wind up unemployed or in prison or both, with the expected negative social consequences. Is that really what you want for your life?

A different approach should be taken for school children with careless task performance attitudes because they're required to be at school irrespective of their wishes. Their careless attitude is because the classes don't keep their interest and they're bored. Sometimes this is caused by exceptionally smart kids who are ahead of the class, and the usual fix for that is a visit to the school to get them to provide extra interesting work for the child. You'll need to do the same thing at home or your kid will get bored with you!

Your Essence Traits

The other reason for a child with a careless task performance attitude is that they can't understand the material being presented and they feel lost in the classroom, so their mind wanders to something they can understand. In this situation the child needs extra help in their school work and, preferably, have work they like to do. For example, a teenager for whom math has always been a mystery and who now finds himself lost in trigonometry might get a better education if that class could be changed to something he likes, such as electronics repair.

The whole subject of education and what should and should not be required for children is continuously a hot social and political topic, everywhere in the world. The consequences of raising children with careless task performance attitudes are under performing individual personalities and an overall negative influence on essence improvement.

Let's consider what's needed to improve a person with an **obsessive** task performance attitude. Obsessive people usually don't realize how they appear to others. Even if some brave soul tells them, they may think this person is jealous of their superiority and trying to sabotage their effectiveness. Has this happened to you? If so, you should do two things:

1. If you're supervising others, stay away from them. That may sound ridiculous, but the meddlesome way you act around them is worse and they need time to calm down. Find out what's going on by regular status reports and, if you don't like it, ask them for suggestions to make things run better.
2. If you're working alone, have someone else monitor you for at least the primary biological needs to keep you alive, such as meals and rest. Don't like having someone baby-sit you? Then find a **determined** person who effectively accomplishes their tasks and copy their schedule. You should find that you get just about as much done and you'll live a lot longer.

If you find yourself missing desirable opportunities and you suspect it's because you have a **casual** task performance attitude, you should make an adjustment. The easiest change you should make, and one that would make the most difference, is to spend more of your time talking about the tasks you do and thoughts on what should be done to do them more efficiently. As a naturally friendly person, your way of expressing your thoughts will not only be received favorably, it'll demonstrate that you have the skills, the ability, and the interest to successfully perform tasks you receive. If you're concerned that you really don't have the skills everyone thinks you have, find someone with a **methodical** task performance attitude and copy some of those skills.

If you have a **determined** task performance attitude and want to become more methodical, spend more time in planning the stages you take to do your tasks. This planning should be written down for two reasons. The first and least important reason is that people will see your methodical brain in action. Hey, a little advertising never hurts. The second and more important reason is that a lot of detail comes between your thoughts of how something should be done and the process of actually doing it. That detail will often be messy and a planning document will help ensure you don't miss important steps.

Control Attitude

The manner in which you try to influence others. Range – **domineering** to **submissive**.

Your Essence, Your Eternity

Image from www.physicsclassroom.com. Used with permission

Evaluation:

Domineering	= 0
Difficult	= 3
Collaborative	= 10
Yielding	= 5
Submissive	= 0

First of all, let's be sure we understand there's a difference between a "take-charge guy" and a guy with a **domineering** control attitude. A person who takes charge of something is generally appreciated because they also take responsibility for its completion, giving the rest of us a break. To be a successful take-charge person means that you have the full support of your colleagues, and doing this requires a **collaborative** control attitude. If you're a domineering person you intimidate your colleagues and often have trouble finding people to work with at all.

Your Essence Traits

A person with a **domineering** control attitude has the following characteristics:

- **Jealous** – when this concerns social relationships it could be exhibited anywhere from unhappy conversations to domestic violence. People who are insecure in their relationships have this characteristic. When jealousy relates to not having something that someone else has, including objects ("Look at the new car Charlie has"), status ("Charlie got that special award"), or career advancement ("Charlie is our new Vice President"). If these topics frequently occur in your conversations and your name isn't Charlie, you may have a jealously issue.

- **Easily angered** – domineering people sometimes use anger to manipulate others because most people try to avoid unnecessary conflict. If you find yourself raising your voice with an angry tone, you may be guilty of this characteristic. Since domineering people often don't know they're being domineering, if people tend to tell you to calm down, think about how you've been acting.

- **Forceful** – if your opinion is asked but you give it to sound like anyone else's opinion doesn't count, you're being domineering. These opinions tend to be black-and-white, with no room for variation or negotiation, and have the effect of intimidating people. You hear this kind of talk from salesmen when the value of their product versus a competitor's product is discussed and from politicians when they're talking about the positions of their opponents.
- **Manipulative** – being manipulative is being forceful in a sneaky way. In this case, you cajole rather than intimidate, but the goal of submission is the same. You sometimes claim that you'll be a victim if they don't do what you want. People do a lot of sighing if you're a manipulative person.

People with a **submissive** control attitude are just the opposite, but just as hard to work with. If you're submissive you shy away from saying what you really believe, especially if someone else has other opinions. You don't voluntarily contribute to activities because you're afraid that anything that goes wrong will be blamed on you. In fact, when something does go wrong, you accept the blame even when it wasn't your fault. While it seems convenient to have someone around to blame, people doing it are showing their own moral and ethical weaknesses, not to mention an inferior control attitude. It's almost a trap for other people, requiring especially strong personality traits to accept their responsibility for failure when you're willing to take the blame. Being submissive hurts you and everyone around you.

You suppress your feelings and repress memories of your being submissive because it's such a negative way to live. You may also hide your feelings of disappointment of not getting what you wanted by implying that your original suggestion wasn't very important to you anyway. You tend to avoid complex activities, claiming they're too difficult. You often wish to be faster, or tougher, or smarter, or more talented in some way, but don't take the necessary steps to do it.

Still not sure if you're submissive? One way is to see how you react to a change in venue, such as someone suggesting going to a different restaurant or movie. If you agree but apologize as part of your agreement, you may be submissive. An example might be, "Sorry, I didn't realize you wanted to go to an Italian restaurant" rather than "Sure, that's OK."

People who are too submissive risk spiraling into even a lower self-esteem, develop internal anger, and possibly more serious psychological problems. Clearly, if you have a submissive control attitude you should take steps to change it.

Interestingly, both **dominant** and **submissive** control attitudes indicate you have an inferiority complex and don't want to be discovered as less capable than your peers. Dominant people bluster their way out of this danger while submissive people shrink away from situations that would demonstrate their inferiority. What's even more curious is that you're probably not inferior to anyone else in the activity – you just think you are and act accordingly. Domineering attitudes show a predator personality while submissive attitudes show a prey personality.

If you have a **difficult** control attitude you're certainly better to work with than if you had a domineering control attitude, but it's not as much of an improvement as it should be. You still get jealous, easily angered, and are forceful and manipulative. The difference is that when you get jealous, it's more of a "Why can't I ever get such good things?"; when you're angered it's expressed as serious impatience, unfortunately too frequently; when you're forceful it's more like being really pushy, but also too frequently; and when you're manipulative it's more easily recognized and ignored. A **difficult** control attitude implies you're frustrated with yourself and are expressing it as if you're frustrated with someone else.

If you have a **yielding** control attitude you accept direction easily, although you may question suggestions before you cave in. Managers like to have you working for them because you don't blindly accept improper direction and your questions, which are always polite, encourage the manager to reconsider their orders. However, you will loyally support the final direction, whatever it is. You don't get jealous, are slow to anger, certainly don't force issues, and can be manipulated, but not easily. You're a very nice person.

The best kind of control attitude to have is one that's **collaborative.** If this is you, you're the most effective because you control not by controlling, but by persuading in a way that your colleagues don't recognize as control.

Nobody likes to be controlled but everyone likes to be included. Many activities are unsuccessful because people involved with them feel they've been excluded from major decisions about the activity or that their opinions/suggestions have been ignored, so they sabotage the efforts. You've probably seen this happen and have felt sorry for the frustrated manager who can't get the activity done. Perhaps you've felt pleased, if you're one of those who have been ignored. In any case, the way to avoid such pitfalls is for a collaborative manager to actively solicit the opinions, suggestions, and help of people who will be part of the activity, and to include that help as part of the project. This is called making the person a *stakeholder* and it works quite well.

You also recognize the help others have given, adding more good feelings for those working on the activity. Of course, if a person's suggestions are wrong for the activity they can't be used. If a demanding or difficult person tries to take control of the activity you have to stop them, but not so they feel personally abused. It takes a lot of patience to understand and work with people with such different personal agendas, but that's what collaborative people do and what makes you have the best control attitude. I never said it was easy.

Your Essence Traits

Improving Your Control Attitude

One of the best self-help books on developing the right control attitude is *How to Win Friends and Influence People* by Dale Carnegie. Following its guidance even slightly will help you improve any control attitude yourself and following it carefully will lead to a collaborative control attitude. Here are the major suggestions from this book:

Fundamental Techniques in Handling People

1. Don't criticize, condemn or complain.
2. Give honest and sincere appreciation.
3. Arouse in the other person an eager want.

Six Ways to Make People Like You

1. Become genuinely interested in other people.
2. Smile.
3. Remember that a man's name is to him the sweetest and most important sound in any language.
4. Be a good listener. Encourage others to talk about themselves.
5. Talk in the terms of the other man's interest.
6. Make the other person feel important and do it sincerely.

Twelve Ways to Win People to Your Way of Thinking

1. Avoid arguments.
2. Show respect for the other person's opinions. Never tell someone they are wrong.
3. If you're wrong, admit it quickly and emphatically.
4. Begin in a friendly way.

Your Essence, Your Eternity

5. Start with questions the other person will answer yes to.
6. Let the other person do the talking.
7. Let the other person feel the idea is his/hers.
8. Try honestly to see things from the other person's point of view.
9. Sympathize with the other person.
10. Appeal to noble motives.
11. Dramatize your ideas.
12. Throw down a challenge.

How to Change People Without Giving Offense or Arousing Resentment

1. Begin with praise and honest appreciation.
2. Call attention to other people's mistakes indirectly.
3. Talk about your own mistakes first.
4. Ask questions instead of directly giving orders.
5. Let the other person save face.
6. Praise every improvement.
7. Give them a fine reputation to live up to.
8. Encourage them by making their faults seem easy to correct.
9. Make the other person happy about doing what you suggest.

Eight Rules For Making your Home Life Happier

1. Don't nag.
2. Don't try to make your partner over.
3. Don't criticize.
4. Give honest appreciation.
5. Pay little attentions.
6. Be courteous.
7. Read a good book on the sexual side of marriage.

8. Listen carefully to what your partner says and make him/her feel important about what s/he says

Vigilance

The extent to which you trust others. Range – unconditional trust to suspicious.

Evaluation:

    Unconditional Trust = 0
    Confident = 7
    Careful = 10
    Skeptical = 7
    Suspicious = 0

Being vigilant sounds like you're distrustful and need to take measures to ensure your safety. It's easy to think "vigilance" and "vigilante" are closely related, but they're not. Vigilante, according to the dictionary, means "one of a group who take upon themselves the unauthorized responsibility of interpreting and acting upon matters of law, public morality, etc." Vigilance is "watchfulness against danger." In defining personality factors, Cattell describes vigilance in terms of the level of trust you have and we'll do the same here.

One of the first things you have to learn as you grow up is how trusting you should be of others. Getting your trust is what everyone wants. Earning other people's trust is something you should want; keeping it requires constant vigilance on your part. Getting and giving trust isn't necessarily based only on how people act – Lisa DeBruine found that people who have similar facial features trusted each other more than people who look different from each other.

Some level of trust is essential for organizations (including governments) to function, so it's important for institutions to foster trust among their members. You may have noticed that when people trust each other their activity runs smoothly, while if people don't trust each other the activity fails. The same is true with governments and you can occasionally see such events in the news. If the leaders of a country don't trust each other, the country operates inefficiently; if they actively distrust each other, the government may fail.

There have been dark times when government leaders have had their own agendas and have tried to destroy, through fear, the trust their citizens have carefully built. For example, in the 1950s the United States suffered a decline in trust when many thousands of Americans were accused of being Communists or Communist sympathizers by the House Committee on Un-American Activities under Senator Joseph McCarthy. Evidence was ignored or exaggerated and certain groups (government workers, entertainers, union activists, and educators) were especially targeted. Many of these innocent people had their careers destroyed and some went to jail. Eventually most of the punishments were declared illegal but it was too late for those affected. Another example was the Chinese Cultural Revolution from 1966-1976, when Chinese leader Mao Zedong declared that "liberal bourgeois" elements were trying to destroy Chinese society by bringing back capitalism. Mao mobilized China's youth to start a class struggle to purge educators and intellectuals from society, and anyone who opposed these practices was considered suspect as well. Many of these former leaders of society were sent to forced labor camps or killed, and much of the intellectual and artistic achievements of China were destroyed.

What is trust? Trust is part of all of the following:
- Trusting that others will respect your feeling and emotions
- Trusting that others won't intentionally hurt you
- Trusting that others won't take advantage of your weaknesses
- Trusting that others will treat you fairly and honestly
- Trusting others to know your secrets
- Trusting others to do what you expect

If you put your trust in someone and that trust is violated, you're less likely to trust that person again. Con artists are skilled at getting people to trust them (to their regret), but the "best" con artists can keep convincing those same people that the difficulty either wasn't their fault or was an anomaly, and if they just keep their faith everything will work out well. The old adage applies here, "Fool me once, shame on you; fool me twice, shame on me." On the other hand, if you've established a level of trust in someone and others tell you that level is wrong, take their advice with caution. You should decide for yourself, through a careful analysis of your relationship, how trusting you should be with anyone.

Your Essence Traits

If you have a **suspicious** level of vigilance you have a tendency to blame or suspect others whenever something looks different, under your assumption that those people are doing something underhanded. If you're immediately suspicious of everyone, even before they've done anything or even said anything, you have a personality problem that needs some attention. You may have developed your attitude through emotional or physical abuse when you were a child. Perhaps you were bullied or belittled when you tried to make relationships, and now your suspicious nature protects you from that happening again. Unfortunately, it works too well because you've detached yourself from society and the few relationships you succeed at developing are rather weak. But your suspicious attitude isn't always bad – suspicious people are highly successful auditors, security investigators, and insurance salesmen.

If you have an **unconditional trust** level of vigilance, I have a bridge in New York City to sell to you. You're excessively gullible and suffer from highs and lows of happiness – the highs when you feel good about helping someone, and the lows when it turns out the person you trusted isn't worthy of your trust. You say that you look at the bright side of others, that you don't want to suspect foul play until it's done because it's just not right to presume evil. Who can argue with that? The problem is that you ignore clear signs that something is going wrong. While you're following the Golden Rule of "Do to others as you would have others do to you", the other person may be following a different rule. Honest people trying to help others must have vigilance while doing it, but honest people with an unconditional trust level of vigilance don't have that vigilance.

Sometimes unconditional trust comes because of your own greed, and an entire industry has sprouted where people are congratulated on winning a lottery they never entered, or getting an inheritance from an unknown relative, or being asked to help someone with a lot of money for a share of it. How much do you trusting people get? Less than zero, because there are always fees or some payment as part of getting your riches. The fees are paid and the riches don't come, but you believe it will all work out. Now, about that bridge ...

If you have a **skeptical** level of vigilance you're very careful when dealing with others but don't immediately think they're up to something. This makes you less likely to be fooled by someone but more likely to think you're being

Your Essence Traits

fooled when you aren't. You generally take a pessimistic view of the situation and "read between the lines" even when someone is talking and there are no lines to read. Being skeptical means your initial reaction is to resist change and to deny agreement. The more enthusiastically a proposal is submitted, the more likely you are to reject it. However, while a suspicious person will reject an idea permanently, you as a skeptical person can be convinced the idea is sound. Most of the reason for you to reject something is the assumed hype that comes from an enthusiastic description. But if that description was from a person you normally trust, ask for more information before you shake your head no. Skepticism has become a fast growing trend because the vast amount of available information with claims and counter-claims makes people wonder who's telling the truth and who's not, and it's safer to be skeptical than it is to be confident in your initial opinion.

If you have a **confident** level of vigilance you tend to "give people the benefit of the doubt" when something negative happens. A good part of this is because you're not burdened by doubts or low self-esteem, are confident in your own judgments, and don't need approval from others. Your

Your Essence, Your Eternity

confidence spills out to others who you try to convince to your way of thinking. If things go wrong, well that's just a temporary setback. You like challenges and competition, work hard, and accept praise with grace. You're even willing to admit your mistakes and use the recognition of those mistakes to improve your future efforts. Sometimes you may be a little too confident in yourself and demand more respect than others are willing to give. Keep working at it – it's only a temporary setback.

If you have a **careful** level of vigilance you combine a healthy skepticism with a healthy confidence in your ability to do it right. You expect to be shown why something is true, not just that it is true; you have to be shown what you're risking by believing someone; you want to see something, not just be told something; and you need to be confident the person will deliver what they promise, when they promise, and for the cost they promise. You're a real leader that others want to follow because you know what you're doing, quickly acknowledge your weaknesses and seek guidance to help them, are as open as possible with your colleagues, and generate a large amount of trust in yourself.

Your Essence Traits

Improving Your Vigilance

If you want to improve any level of vigilance, try communicating more. A high level of communication is the most important aspect of people with a **careful** level of vigilance and is lacking by degrees in people evaluated at other levels. Suspicious people complain rather than communicate, when they share their feelings at all. Unconditionally trusting people accept the situations rather than communicate their concerns about them. Skeptical people need to ask for information to assuage their fears. If the description was from a stranger you'll need even more information, but that's up to you. If you're skeptical and don't communicate the other person will think you're just being stubborn and your vigilance level will remain as skeptical as it was before. Confident people communicate their confidence but also need to receive information about possible problems. That's the main difference between a confident person and a careful person – the ability to extract information on problems so they can be addressed before they occur.

Here are qualities you should consider to improve the level of trust in your life:

Your Essence, Your Eternity

- **Belief that mankind is essentially good.** Yes, there are criminals and terrorists lurking around, but if you view people in general as potential criminals and potential terrorists, your vigilance concerns will be too high and your relationships will suffer. Your friends are very much like you, so if you trust yourself you should be ready to trust them.
- **Belief that life is essentially fair.** Yes, it's easy to find bad things happening to good people and good things happening to bad people. The media dramatizes such events to attract viewers. But overall, the people you work with and play with are as fair as you are.
- **Focus on fixing problems rather than blaming individuals.** Being vigilant isn't finding the person to blame, it's finding problems before they occur and stopping them from happening. If you're known for blaming people, those people will lose trust in you and your overall ability to perform will be reduced.
- **Admit your own mistakes.** Everybody makes mistakes, and when you take responsibility for yours other people will believe that you'll be as trustworthy

in the future. Failure in competence can be forgiven, but failure in admitting your failure in competence can't. Only people with immature personalities never admit they're at fault.

- **Reduce your fears**. We all have fear of failure, fear of being hurt, even fear of the unknown. It's easy to imagine worst-case scenarios, but if your life runs on protecting yourself from them then other people will see you as difficult. Success in every activity comes with the admonition, "Don't freeze!" If you freeze when you're taking a test in school you won't remember answers you know; if you freeze when you're talking with someone you hope will like you, it's much less likely they will. If freezing in difficult situations is something you've noticed in yourself, work very hard at overcoming it before you take that first parachute jump.

Dependability

The extent upon which others can rely on you. Range – **never** to **always**.

Your Essence, Your Eternity

Evaluation:

Never	= 0
Rarely	= 2
Sometimes	= 5
Usually	= 8
Always	= 10

Dependable people are always desired, whether for work or for play, no matter what the situation. If you're a surgeon you want dependable people by your side; if you're a thief you want a dependable lookout.

You should want to be considered dependable by everyone you meet and know. If you don't want that, it's because you're trying to get out of doing something and hope the other people will pick someone else to do it. This attitude has an undesirable longer-term effect in that those people will project your undependability onto other activities that you really want to do. Too bad. Being dependable is something you are or you aren't, and the lowest standard you set will probably be the standard you live with.

Characteristics often used to measure dependability are:
- Do you show up on time?
- When you say something is going to happen, does it happen?
- When you describe something and someone else investigates it, does the original description match what you've said?
- Do you lie to people?
- Do you help other people lie?
- Do you withhold information to avoid conflict?

The evaluation criteria will obviously be different for different people. Your estimation of your own dependability will be different from someone else's estimation of their dependability, even if you're both equally dependable. That's not a problem – your goal is to be always dependable, so your rating now will change as you improve.

Improving Your Dependability

Here are some qualities you can focus on to make yourself more dependable:
- **Be observant**. Understand what it is about you that makes you less dependable than you'd like to be.

Your Essence, Your Eternity

- **Be time conscious**. People hate waiting for you to do something, and the more reliable you are in meeting your time promises the more dependable you'll be to them.
- **Be honest**. Tell the truth and admit your mistakes. Nobody's perfect, and when others see that you recognize deficiencies in yourself they'll see you as a more dependable person.
- **Be diligent**. By meeting the goals you promise you'll demonstrate how dependable you are. One trick to this is to promise a little less than you think you can do, because problems always come up and addressing those problems will take time from the planned work you have. By promising less you're giving yourself time to resolve those problems.
- **Be consistent**. People get nervous when they see someone acting inconsistently and think that person doesn't know enough about the task to dependably accomplish it. Yes, consistency is boring but so is dependability, and you're trying to be more dependable.

Your Essence Traits

Extroversion

Level of Engagement with the External World

Warmth

The extent of your attention to others. Range: **very distant** to **very outgoing**.

Evaluation:

Very distant	=	0
Partially distant	=	3
Neutral	=	5
Partially outgoing	=	8
Very outgoing	=	10

Evaluating your warmth factor can be tricky because you have to identify your true warmth rather than your fake warmth. Are you a politician or a salesman? Then you're used to showing an outgoing personality to everyone, irrespective of how much of a bore you think they are. Do you play golf because people expect you to play it, even though you don't really like the game? When you evaluate your warmth, strip away these pretensions and look at yourself with a cold light.

When you meet a person the first impression you have is based on their physical appearance. As you know, physical appearance is a shallow assessment of the person, sometimes described with words such as "trophy", "fox", and "eye-candy" for physically attractive people, and "dog", "hag", and "gargoyle" for those who are beauty-challenged. The real attractiveness in a person is the kind of person they are, in other words, their essence.

One of the first things you notice about a person, and one of the most important, is how important you seem to be to them. Everyone likes to feel that they're worth your attention, and no one likes to be ignored. The evaluation of your warmth is a reflection of how your attention to them is perceived. If you tend to ignore people, you're evaluated as **very distant** and get a very distant zero. The less distant you seem to be, the higher your warmth evaluation becomes, reaching the top score when you're perceived as **very outgoing**.

There are confounding factors that affect perceived personal warmth, factors that you should recognize when interacting with others. For example, some people are naturally shy until they become better acquainted; some people may have communication difficulties (such as speech defects) that inhibit them from expressing their true natures; some people may be afflicted with diseases, such as Down's syndrome, that cause them to be less outgoing, or rage disorder, that cause them to be *too* outgoing; some people may be outgoing only because they want something from you. We're all familiar with how attentive some politicians are during election campaigns, then you're ignored until the next campaign.

If you had a choice, would you want a warm personality or a cool one? In this case, "cool" means "not warm" rather than the slang definitions of "excellent" or "great". Warm personalities generally attract more friends, while cool (not cold) personalities are perceived as more independent. People with warm personalities are considered affectionate, caring, and tender. They're also viewed as more easily manipulated because they're affectionate, caring, and tender. People with cool personalities can certainly be affectionate, caring, and tender, but they don't exhibit those characteristics as prominently as people with warm personalities.

Your Essence, Your Eternity

In general, a warm personality is desirable because it connects you more closely with your environment – including people, cultures, and habitats. Without these connections you will likely be less interested in your surroundings, resulting in greater tensions with your own culture and habitat.

Improving Your Warmth.

There are two time periods when you can improve your attention to others – when you're with them and when you're not. Being "with" someone includes any kind of direct contact, whether it's physical or remote, real-time or at separate times (such as when someone is reading a letter you wrote). Direct contact is by far the most common, and most of this section will address how you can improve that aspect of your warmth.

You still have a warmth factor of your personality when you're not directly interacting with anyone at all. This happens when you send an anonymous gift to a worthy cause. Churches that pass collection plates tap into this form of personal warmth; those that automatically charge your credit card for a specific amount every month do not. If you get a receipt for your gift, it's not anonymous. Note that if

Your Essence Traits

you gain any benefit from your gift, such as claiming a charitable donation on your income tax, it's no longer anonymous. Giving with no recognition is the highest form of personal warmth and the hardest to do; almost all of us want to be recognized for the good deeds we do. But you do have the satisfaction of knowing that other people were helped. To improve your ability to give at this high level, start by sending a small amount to your favorite charitable cause, specifying it as anonymous. If you feel good about it, you're improving your personal warmth.

Now let's look at direct contact. Too often, giving your attention to others occurs when you think about another person to help them in some situation. There's certainly nothing wrong with that when your assistance is requested, but many times you try to help anyway. When you do that, you're not really focusing on the other person; you're focusing on your critical analysis of their situation and your thoughts on how to improve it. In other words, you're being a busybody. What you should do is put aside your criticisms and judgments to focus on what the person is asking of you.

Sometimes there are other factors affecting your warmth with others, such as:

- **People with diseases affecting their personality**: Some diseases that include physical body characteristics are easy to see. Other diseases that affect the brain and express themselves with personality disorders can be hidden until you blunder into them by talking to the person. Why was he so rude? This is where your personal warmth can really shine.

  First, of course, you have to be sure it's a disease that's causing the poor conduct. It could be that the person is just nasty. If you decide to find out by asking them, give yourself a zero for warmth and a lower evaluation for maturity while you're at it. But it turns out you don't have to know at this initial meeting. Patience is the best approach in these cases, just as it was with people with communication difficulties, but you have to be extra patient because you don't know if/when it will work. If patience does work before they drive you crazy, congratulations and give yourself a 10 for warmth. If you don't seem to be making headway no matter how pleasant you are, make a polite exit. Sometimes all you can do is try your best and, if you do, you have good personal warmth.

- **People with agendas**: Sometimes you meet a person with tremendous personal warmth who gives you all the attention you desire. You may think "This is a wonderful person – I'm so glad we met" and you may be right. However, the person may be actively seeking your attention to sell you something, to get you to support something, or even because they intend to swindle you. While extending your personal warmth to people is a positive personality trait, you still should evaluate the people with whom you're associating. Don't reject someone just because they're being especially friendly – you'll lose a potentially great friend. But if your new friend starts asking you for things, look closer. You may want to adjust your personality to be a bit cooler with them.
- **Non-people**: Non-human living things that normally have close associations with humans will appreciate your personal warmth even more than some people do. Animal lovers know that the more attention you give to animals in your care the more responsive and better behaved those animals are. Wonderful

relationships can be developed with other species by creating trust through your personal warmth. You have to use common sense, though – wild animals are wild animals, no matter how much personal attention you give them. You should only project your personal warmth to domestic animals; if you do it with the bear you just met in the woods you'll probably regret it. Scientists who interact closely with animals in the wild have learned over time what to do and what not to do. You can't charm a snake with your personality. Stick to your dog or cat, and good luck with the cat.

Privateness

The ease with which people can get to know you. Range – **extremely private** to **extremely open.**

Public Toilet with One-Way Glass Walls

Your Essence Traits

Evaluation:

Extremely private	=	0
Private	=	5
Cautious	=	10
Open	=	5
Extremely open	=	0

Have a lot of friends? If you don't, part of the reason may be that you keep yourself distant from people until you know enough about them to be comfortable. While you have fewer friends, those you have are just the kind of people you like. You might be envious of others who make friends easily, but you don't feel that bad about it because you're sure those friends are just superficial, drawn to a bubbly person who likes to talk.

You're wrong, you know. People are naturally attracted to someone who seems interested in them and wants to connect (see the discussions on warmth and sensitivity). Part of that connection is developing a relationship in which both parties understand more about each other. Relationships are superficial if they're based solely on physical attraction. Those that are based on compatible personalities are friendships that last, and the only way to know if your personalities are compatible is to talk to each other.

But if the other person is so open that you hear things you don't want to know, the relationship will be short lived. Do you really want to hear about his hemorrhoid operation? If the discussion turns to taxes (always a sensitive topic) and she asks about the deductions you claim, do you tell her or run for the door? You can sometimes recognize if someone has an extremely open approach to their privateness if they start gossiping about others. Gossips have push-pull relationships with most people – you're pulled to them because you're interested in the latest dirt, but you're pushed away because you don't want your life displayed to the world. If you choose an extremely open person as a friend, you take your chances.

Except for the extremes, a degree of privacy is a positive trait in a personality. At the same time, some openness is needed to show others you're willing to share with them. If you're more private than open you can do well in occupations that require some secrecy, such as a diplomat or a personnel officer. If you're more open than private you're better in marketing and artistic areas. Of course, there are times when an open person should be private and times when a private person should be open. For example, say you and your girlfriend are suspicious types so you've agreed not to keep secrets from each other. But now your brother tells you something that he wants you to keep just between you and him. Should you tell your girlfriend? If you do, you're breaking a trust. If you don't and she finds out, you're toast. What you do is tell her that your brother asked you to keep something secret so you can't tell it to her. If she becomes upset, say that you think your boyfriend-girlfriend relationship is strong enough that you wanted her to know you're keeping a secret. This approach keeps your integrity with both your brother and your girlfriend, and both should respect you for it. This same approach should work with a variety of relationships, except if you're a spy and are being interrogated by the enemy.

Improving Your Privateness

Privateness is an attribute in which good judgment is the key to how private or how open you need to be in each situation. There's no right or wrong level of privateness that's always correct for you. If you're naturally private you'll feel uncomfortably exposed if you try to be too open; if you're naturally open you'll feel like you're hiding something if you try to be too private. Pick a level that's close to the center but that's still within your comfort zone.

If you think you're too private when you should be more open, pick the least private part of the situation and include it in your discussions. Then pick the next-least-private part and start talking about it. Continue in this fashion until you begin to feel that you're saying too much, or you see your audience isn't connecting with you as you'd hoped, or you've reached a place where private really means private. Don't gossip.

If you think you're too open when you should be more private, quickly change the subject and try to pass the conversation to someone else. If your audience returns the subject to you (most people like to hear gossip, after all), suddenly discover that you don't remember any more about it. They may not believe you've said all you know, but if you just stop talking there's little they can do that's not rude and they'll have more respect for you. As you become better at restricting what you say in situations that deserve privacy, you'll feel better about yourself and your friends will begin to trust you with private matters of their own.

Everyone has situations where they can't seem to stop talking, even though they absolutely want to, but some people seem to be addicted to talking. If you really can't stop talking, no matter how hard you try, you might consider professional help – your personality will be better for it.

Material Attitude

The extent that material possessions are important to you. Range - miserly to wasteful.

$1,000,000 in $1 Notes
Bureau of Engraving and Printing, Washington, D.C

Evaluation :

Miserly	=	0
Frugal	=	5
Careful	=	10
Consumer	=	5
Wasteful	=	0

Your Essence, Your Eternity

Are you economical or extravagant? Do you save for the future or spend what you have now? Your material attitude is not something as natural as being moral or ethical; it's something that's learned from your childhood and further learned as you go through life.

People brought up in hard economic times, when everyone is poor, are careful with their spending as they become more affluent. However, if they attain a level that makes them feel wealthy, these people want to flaunt it a little as a kind of self-justification to show they've made it. Not too much, though, because they're still afraid of being poor again. If they become quite wealthy their philanthropic switch turns on, allowing them to further demonstrate their high economic status by actually giving their money away. To people who were raised poor, giving money away is the ultimate expression of wealth. Other people, with less of a need for public recognition give anonymously, which has the double benefit of helping others while giving the individual a sense of personal pride. And, of course, there are always those people who are just greedy. A popular bumper sticker in the 1980s to characterize the excessive greed at that time read, "He who dies with the most toys wins". The New Testament disagrees – "For what does it profit a man if he shall gain the world, yet lose his own soul." (Mark 8:36)

Encouraging people to desire material goods is an entire academic discipline and commercial enterprise called "marketing". The idea is to convince people to buy something they really don't need by showing how special they'll be if they have it. Don't you want to be "better" than your annoying neighbor who just bought that other product? Perhaps it's important to use the same after-shave as your favorite athlete (unless, of course, you're a woman). You think that in some magical way the qualities of the celebrity will be transferred to you. When they aren't, you try another product. One of them has to work, doesn't it?

Some people see the process of attaining wealth as a hollow achievement that forces them to work harder and harder at tasks they dislike and deprives them of fulfillment. These are the people who talk with increasing disgust about the "rat race" and who take one of two divergent paths: most of them feel compelled to keep up with or exceed the material possessions of their peers (cars, vacations, special schooling for their children, etc.) and who continue their quest for the financial means to achieve their perceived needs. The other, much smaller number of people, opt out of the race and live an "alternative" lifestyle.

There are traps on both sides of this material attitude road. If you strive to "keep up with the Joneses" you have to work long hours and have both adults in a family employed, sometimes with multiple jobs. Extra expenses, such as childcare, are involved and the not-so-hidden effect of having someone else, with their own perspectives on life, bringing up your children. And you may still not make enough money! Even if you do, you may now try to move up a notch to a more affluent set of friends, which reinvigorates the need for further money for your higher cost material needs. If this is your path, be sure to set aside money for psychiatrists.

On the other hand, people who avoid monetary stresses and have fewer material possessions, more family time, and more time for reflection have their own issues to consider. A few of these people completely opt out of society by becoming subsistence farmers or hunters living in a remote cabin. Those are escapes rather than attitudes and should be addressed as unfulfilled psychological needs rather than as material attitudes.

Your Essence Traits

We can disregard the fact that less money will mean you'll have fewer educational opportunities (which aren't material possessions), but you may not be able to afford those expensive hiking shoes, or the ski equipment, or the lessons from that famous professional. That means the experiences you do have will not be as substantial, and without the right equipment and instruction your chances of really excelling in an activity are remote. Similarly, if the latest Internet access is beyond your budget you'll have to accept that much of the world's knowledge will also be beyond you. You have to decide what level of material possessions makes sense for you, and that decision will determine your material attitude.

Very few people, fortunately, are so obsessively frugal that they're considered **misers**. These people get most of what they have by buying it second-hand (or worse) and never buy anything new if they can avoid it. They're often as affluent as you are but just can't bear to spend their money. You might see someone selling used underwear and wonder who would buy such a thing. Now you know. If you visit a friend's house and notice that the bar of soap in the bathroom is one of those tiny soaps from hotels, and that it's been used for awhile, your friend is a miser. People with this level of material attitude are more than cheap – they have a psychological problem that's inhibiting them from having the fulfilled life that's available to them.

A good (or bad) example of a miser was Hetty Green (1834-1916), said to be the richest woman in the world in the early 1900s, worth perhaps $200 million when she died. It was reported that she never used heat or hot water because of the cost and that she wore clothes until they wore out before she changed them. When her son broke his leg she took him to the charity ward at the hospital but was recognized and told she was too wealthy to use charity services, so she decided to treat his leg herself. It contracted gangrene and had to be amputated. If you become a terrible miser like this you also may wind up being vilified in a future edition of this book.

While nobody loves a miser, nobody loves a **wasteful** person, either. You're obviously wasteful if you usually buy more food than you'll use and don't mind throwing the rest away. You're obviously wasteful if you wash your car during a drought when you've been asked to conserve water. In cases like those you exhibit a level of selfishness and greed that demonstrates an uncaring attitude towards your habitat and culture that's expressed in your material attitude.

Your Essence Traits

But some people consider you wasteful if you buy a newspaper rather than read the news online because of the environmental resources needed to produce the newspaper. If you don't drive a car designed for lower emissions, you might be called wasteful. Further thought must be taken when identifying a person as wasteful in such circumstances. If you didn't care about wasting resources or about polluting the environment you'd be wasteful but otherwise, from a material attitude perspective, you're merely a consumer. In fact, if you were actively trying to lower your negative impact on the habitat you'd be both frugal and congratulated. Keep up the good work!

Saying a person is frugal is a polite way of calling him cheap. Frugal people are the ones who line up for sales and look to the price as more important than the features of a product. When a frugal person buys a car they drive it until it's more expensive to fix than to buy another one. They attend performances when the ticket price is lowest and they bring their own snacks. They prefer to travel the least expensive way possible, even if it's considerably less comfortable and takes more time. However, frugal people are often very knowledgeable about where and when to find bargains, so it's in your best interest to have a frugal person help you find what you need. If you're not a frugal person yourself you don't want them to order for you in a restaurant; when you open a present given to you by a frugal person, thank them anyway.

On the other side of the material attitude center is the **consumer**, a person who looks for new products or features and who likes to shop for them. If this sounds familiar, you're the focus of the advertising industry and one of the people who determines whether an economy is healthy or not. You follow fashion trends, buy new cars regularly, stay at tourist hotels, and eat out at least once a week. You sometimes get into trouble from overuse of your credit, and you're the successful target of fraudulent proposals. You may not be the best shopper (just because everyone's buying a brand doesn't mean it's the best brand) but you shop frequently. Because you have a relatively high need for material items, you sometimes "shop 'til you drop" and wind up with a lot of little-used merchandise. Do you own 20 pairs of shoes? You're a consumer. Does your children's bedroom look like a toy store blew up? Not only are you a consumer, you're training the next generation of consumers.

Your Essence Traits

People with a **careful** material attitude combine the best parts of being frugal with the best parts of acquiring material possessions. You have a careful material attitude if you acquire only what you need through a review of available products. It's not necessary to spend a lot of time pouring over advertisements and product reviews to be a careful person; you just need to find good sources of information to help you understand your options. You don't overspend your credit limit and you pay your bills before additional fees are added. Having a careful attitude towards material possessions benefits you in other ways, too, because you're happy with the lifestyle you've fashioned and you've created a peaceful barrier between yourself and those who insist you acquire more.

In general, a superior personality considers material possessions to be secondary to personal interactions and experiences. If you love material possessions you should keep in mind that they'll never love you back.

Improving Your Material Attitude

If you're a **miser** and you don't improve your material attitude your life will be friendless and the only satisfaction you'll have is that you spend very little money. You know you're not liked by anyone and if that makes you happy, a miser you'll stay. But if you want to develop relationships with people, you'll have to recognize that you have this personal issue. As mentioned above, a miserly attitude is a psychological problem that should be analyzed by a professional. The problem, of course, is that you're a miser and won't spend money on a psychiatrist. Don't take the cheap alternative of getting advice from your brother-in-law who had two psychology courses in college because that has a better chance of making you worse. One approach might be to go shopping with a frugal person you trust and follow their advice on what to buy. The idea is to gradually loosen the grip your stinginess has on you so you can begin to join your culture. Relax and take your time with this process – you don't want to suffer any worse psychological problems than you have right now.

If you're **wasteful** you're berated by everyone except other wasteful people about the misuse of natural resources, and you have no defense. Fortunately, you also have an easy fix – stop being wasteful! If something has an expiration date, be sure you don't buy so much that the remainder will be expired. If you're asked to conserve something because shortages exist, do it. While this may seem so obvious it shouldn't have to be said, wasteful people have a natural level of selfishness and these situations are not obvious to them. The real correction for a wasteful person is to develop an appreciation for other people's needs and to adjust your own needs, as sensibly as you can, to accommodate them. To help your wasteful material attitude, work on your warmth, sensitivity, and fairness personality attributes.

If you're not one of the hopeless misers or wastefuls, one approach to improve your material attitude is to develop a feeling of gratitude for what you've got. In his research on gratitude, Robert Emmons had the participants keep gratitude journals in which they reflected on the positive aspects of their lives, and just the act of doing those reflections made them more satisfied with what they had. These people felt less disappointment, disillusionment, regret and frustration about their material prosperity, and they reported feeling closer and more connected to others.

Consequently, people who have **frugal** or **consumer** material attitudes can incorporate grateful feelings while moving towards **careful** material attitude. If you're frugal, use your frugality to find the best buys on new and improved products. Allocate special times (birthday, holiday, etc.) to give yourself a treat and buy something to replace an older item you would normally keep. If you're a consumer, say no to every second or third new item you want to buy. This doesn't mean you decide to buy 33%-50% more items so you can turn them down – you have to be honest with yourself to improve yourself. The idea, in all cases, is to develop a material attitude that allows you to get what you need without having the possessions give the impression you're stingy or that those possessions run your life.

Risk Attitude

The extent of your willingness to take chances. Range – extremely adverse to reckless.

Your Essence Traits

Evaluation:

Extremely Adverse	= 0
Guarded	= 7
Neutral	= 10
Seeking	= 7
Reckless	= 0

When we talk about risk attitude we have to be specific about what kinds of risk we mean. The general types are

- **Risk of having an accident**
    - Do you climb on the roof to clear leaves from the drains or do you call Jim the Roof Man?

- **Risk of getting an illness**
    - Do you get a flu shot every year or do you take your chances?

- **Risk in recreation**
    - Do you try that advanced ski run or do you stick to the Bunny Slope?

- **Risk in social settings**
    - Did you ask that pretty girl for a date or not?

- **Risk to your financial situation**
    - Do you prefer the stock market or a bank savings account?

In research, this is called "domain-specific risk attitude." You've certainly seen some people (perhaps yourself) become hypersensitive to some kinds of risk while being seemingly nonchalant to others. Certain occupations, such as policeman and fireman, while inherently dangerous, are admired by society as important to safeguard the welfare of everyone. Other high-risk occupations, such as mining, may be the only jobs available to a person. Some risk attitudes are personal perception, others are received from the outside, and others are a combination. For example, you may personally think mountain climbing is just too darn dangerous, or your mother lost her brother in a mountain climbing accident and now you're not allowed more than 20 feet above sea level, or none of your friends like mountain climbing so you don't do it, either.

James Byrnes and his colleagues analyzed a variety of risk studies and found that, in general, men take greater risks than women. The difference depends on what the risk is: drinking and drug use were about the same but men were riskier drivers.

With all this in mind, when you reflect on your risk attitude you should consider each type of risk individually and not try to form a general impression of your average risk attitude. Averages can be misleading – if every year you live six months in Maine and six months in Florida, on the average you live in Virginia. To avoid such erroneous conclusions, your risk attitude evaluation should be done individually for attitudes towards accidents, illnesses, recreation, social settings, and financial situations.

When your situation appears to be insecure, you look for more security. This is true on a personal level, when you choose to take that boring government job because it provides employment security. Insurance companies encourage insecure feelings in people to get them to buy more insurance. As with all other situations in life, your opinion for or against insurance is an indicator of your risk attitude.

An **extremely adverse** risk attitude means you're not willing to take any risks at all. This risk attitude is a reflection of a very pessimistic person with no confidence in your abilities. At the extreme you'll suffer from *panaphobia* (spelling varies) which is fear of everything, including a fear that you've spelled misspelled panaphobia. Besides the obviously dreary life you'll lead, accomplishing almost nothing because you're too scared to try, an extremely adverse risk attitude is self-fulfilling. Just because you're sure bad things will happen, they will. If you're afraid you'll get motion sickness, you will. If you're afraid you'll double fault your tennis serve, you will. If you're afraid you'll get a zero for having this extremely adverse risk attitude, you will.

On the opposite end, also with a zero evaluation, are people with a **reckless** risk attitude. You never consider the risks in what you do, even when the consequences are possibly fatal. While the *Darwin Effect* (in which a person does something so stupid they get killed, keeping their genes from creating a new generation of idiots) will usually take care of things, sometimes your activities encourage others to try their stunts. In a probably vain attempt to curtail such copycats, recall what Evel Knievel, the famous motorcycle daredevil, said:

> There are a lot of myths about my injuries. They say I have broken every bone in my body. Not true. But I have broken 35 bones. I had surgery 14 times to pin and plate. I shattered my pelvis. I forget all of the things that have broke.

A reckless risk attitude also appears in the other types of risks, such as when you invest in proposals that promise great wealth in a very short time, only to see your money

Your Essence Traits

vanish without a trace. While sometimes you learn from your first experience and never repeat this behavior, usually recklessness is part of your personality and you wind up following "if at first you don't succeed, try, try again." A well-known example is Sir Isaac Newton who got caught up in buying shares in England's hottest stock, the South Seas Company, that was establishing a shipping operation. He wound up losing his life's savings and for the rest of his life would not permit anyone to speak the words "South Sea" in his presence. A quote attributed to him is "I can calculate the motions of the heavenly bodies, but not the madness of the people."

A **guarded** risk attitude is an indication of a careful person. If you always make sure your house is secure before going to bed, if you always come to a full stop at stop signs, if you take your vitamins every day, if you dabble in the stock market (but not too much), you have a guarded risk attitude. This doesn't sound bad, and it's not as long as you restrict your attitude to yourself. When you start advising/directing others, though, you become a nag (or worse). "Make sure you brush your teeth after every meal", "You're driving too close to that car", "You shouldn't take such chances with your money" and similar remarks drive other people away. Of course, you're only trying to make their life better and can't understand why they don't appreciate your help. We've all been guilty of this at one time or another, especially when we really do have more knowledge or experience in the activity. Enjoy your quiet life.

A **seeking** risk attitude, on the other hand, applies to people who enjoy taking chances, but not dangerous chances. Perhaps you'd prefer to be called an "adventurer", and that's also a fair description of your risk attitude. You're a heavy investor in the stock market, but you do watch the trends so you don't end up homeless. You enjoy pitting your skills against nature and look forward to that trip up the Amazon and that camping trip in the backwoods. You enjoy competitive sports, often more for the competition than for the exercise. If a tornado watch is sounded, you're one of the watchers while your friend with the guarded risk attitude is packing supplies in the basement. You should also restrict your attitude to yourself, because other people may be uncomfortable with the risks you're willing to take. Sneering and calling them "sissy" or "wimp" won't help your social status, either. Enjoy your adventurous life.

Your Essence Traits

A **neutral** risk attitude combines the best of the guarded and seeking attitudes. You look for good investments for your money, but keep enough in a safe bank account in case those good investments go bad. You look forward to that backwoods camping trip, but make sure you have the vitamins you take every day, adequate medical supplies, and a way to communicate for help. You're adventurous with seeking people and quiet with guarded people, so your social calendar is probably full. Enjoy your life!

Improving Your Risk Attitude

Here are ways for you to develop a neutral risk attitude for each of the major risk types mentioned above.

Risk of having an accident: The major causes of accidents, in order, are carelessness, inexperience, and having the wrong attitude.

To avoid carelessness, try to be alert but not obsessively alert unless, of course, you're in a dangerous situation. Carelessness happens when you take a hazardous shortcut or when you get too tired to pay attention. Have fun, but don't be careless when you're doing it.

Accidents caused by inexperience generally occur when you're trying to operate an unfamiliar piece of complicated equipment or when you're trying to perform an unfamiliar maneuver. If you've never used a table saw, don't try it without at least reading the instructions. You don't have to become an expert with the equipment, but you do need to know important aspects of its operation and then be appropriately terrified while it's running. If you've enjoyed using a body board in a one-meter surf, don't think you can do the same thing in a ten-meter surf. You know that the more experience you have at any activity the better you are at it. Give yourself the time to get that experience.

Having the wrong attitude is a prescription to having an accident. You may start out understanding and appreciating the risks of an activity, but over time you become complacent and are less likely to follow precautions. The adage, "There are old electricians and there are bold electricians, but there are no old, bold electricians" applies. Just because you've never had a car accident doesn't mean you can stop wearing your seat belt. Yes, experience can change your initial strict attitude to one that's less restrictive, but consider the reasons you have to change and recognize the possible effects of that change.

Risk of getting an illness: Germs are all around us all the time but we remain generally healthy because our body's immune system keeps out the bad guys. But sometimes our immune system is compromised, and five reasons for that are initial contact, genetic influences, external influences, psychological influences, and your lifestyle.

Initial contact and genetic diseases are something you were born with. The difference is that if there's no obvious family history for a disease you were born with, it's called initial contact; diseases that have occurred as part of your family's history are called genetic. Sometimes you don't realize you have a problem until it suddenly appears. Examples are some forms of childhood cancer, infantile diabetes, and Down's syndrome. There's no risk of getting the illness – you already have it – and all you can do is to get the best medical attention you can and to fashion your life to make the best of it. In this case attitude is especially important, because it's easy to become bitter at the situation you have in comparison with those around you, but living your life as a bitter person is a waste of your life. You can find people with illnesses worse than yours who have become shining examples for all of us to follow. That's the kind of essence to strive for, ill or nil.

External influences are disease density, temperature and humidity. If everyone around you is coughing, you're more likely to get sick. Since high disease density means a higher occurrence of transmitted infections, isolating sick people (in extreme cases, quarantine) or removing yourself from populated places may be the best ideas. During the Black Plague in Florence in 1348, Giovanni Boccaccio wrote *The Decameron*, a story of seven men and three women who escape the disease by fleeing to a villa outside the city.

If you're too cold or too hot your body focuses on relieving that stress, allowing a possible entry for germs. If the air is too dry your nasal passages may develop cracks, allowing germs to creep in. If the air is too damp that dampness itself is a breeding ground for germs that may be able to overwhelm your initial defenses. What to do? Pay attention to how your body feels and make appropriate corrections. There's no manliness in going outside wearing shorts in freezing weather, but there will be a lot of head-shaking from your family if you come back sneezing.

Psychological influences, especially stress, are a common cause of surprise illnesses. People are often surprised to learn that all stresses, including "good" ones, reduce the ability of your immune system to fight off germs. Let's say you just graduated from college, got married to the man of your dreams, started a wonderful career, and bought a perfect house. Boy, are you in trouble! Each one of those events is a major life stress and you're much more likely than you think to get sick. Sometimes the stresses will cause a life-threatening illness like a heart attack, even in relatively young people. Make note of the stresses you've encountered and pay careful attention to how your body feels.

Of course, your lifestyle plays a large part in how prone you are to illnesses. Do you smoke? Better be prepared to cough up a lot of money for medical bills. Do you drink excessively, are frequently and intensely aggravated at people or things, have a poor diet, or get very little sleep? You're asking for trouble. The good news is that lifestyle changes are the easiest to make and make the most difference when you do. Review the kind if lifestyle you lead and decide if changes should be made. When you've made the changes you should, you know the care you take will improve your chances to have a long, healthy, and happy life.

Risk in recreation: We're not talking about extreme sports here, which are really performance events suitable only to professionals with extensive training. Let's consider recreational activities normally enjoyed by many people and the risks to consider. Do you know what's considered the most dangerous recreational activity by the National Center for Catastrophic Sports Injury Research? Cheerleading, which has accounted for over 65% of catastrophic sports injuries among high school and college girls for the past 25 years. Other dangerous activities (in order) are horseback riding, lacrosse, gymnastics, bull riding, soccer, motocross, hockey, football, and rugby. You're probably surprised that some activities, such as hunting, aren't on the list. That's because there are fewer serious injuries reported, although they do make the front page when they occur.

All recreational activities have risks and you have to decide if those risks are suitable for your own risk attitude. Whatever activity you choose to pursue, take the suggested precautions against serious injury. Then you'll be able to enjoy yourself as the activity was designed to do.

Risk in social settings: Most of the risks in social settings revolve around sex. We start with the fear of being rejected – ladies being rejected by not being asked for dates and men being rejected by the lady they just asked. The best antidote to fear of rejection is projection of self-confidence. You know you're a wonderful person with wonderful qualities, and if someone else is too blind to see them, that's their problem. Self-confidence doesn't mean haughtiness though, so take it easy.

Besides rejection, the dangers of casual sexual activity have been discussed and documented under so many subjects that we won't repeat them here, except to say that the risks include physical, psychological, emotional, and financial problems for those involved. These same issues are present throughout dating activities and if you think you may be a victim, discuss it with your partner and with people you trust before you take your own actions. Communication is the key to any social relationship and misunderstandings easily occur without it.

Risk in other social settings concerns the specific group involved. You may be afraid you'll do or say something very inappropriate for your audience, but you probably have very little to worry about. The group invited you to join them, so just be yourself and enjoy the experience. This is true whether we're talking about a White House dinner or a motorcycle gang beer blast. You were invited because they thought you belonged.

Risk to your financial situation: Money management is a very personal activity. If you're a natural risk taker, take your risks but be prepared for the consequences. No whining. However, you might take a bit of your risked money and put it somewhere safer, just in case. History is loaded with examples of people rich on paper with their investments, but then something happened and all their money was lost. You probably know someone just like that.

If you can't bear the thought of losing your money in an investment, put it in a bank savings account and sleep securely every night. But when your risky friends tell you how well their investments are doing you have to accept the pitiful return your money is getting. No whining. However, you might take a little bit of your money and invest it in something more risky, just to see what happens. You have to understand that this money may be lost, but doing this will put a bit of adventure in your otherwise dull financial life.

Leadership

The extent of your interaction in a group. Range – **independent** to **both follower and leader.**

Your Essence Traits

Evaluation:

Independent	= 2
Follower	= 7
Leader	= 7
Both follower and leader	= 10

When you're interacting with others, it's good to know your place. Animals do this by a wannabe challenging the leader in the hopes of becoming the new leader. Humans sometimes do the same thing, most visibly in sports, but usually each person's place in a group is decided by the group because of the person's ability and personality.

As you can see in the evaluation scoring, **followers** have the same score as **leaders**. This is because both leaders and followers are equally important components of every group. An activity will fail if everyone in it is a leader who wants everyone else to follow them, and it will fail if everyone in it is a follower looking to be led. If you don't like to lead and don't like to follow, in other words you're an **independent**, you're more of a maintenance challenge than a useful contributor to the group activities. Thomas Paine, a prominent figure in the American Revolution, said "Lead, follow, or get out of the way." The best leadership attribute is when you can both lead and follow, depending on your station at the time.

Followers follow instructions. The only skill you need to be a follower is to recognize that someone else is giving those instructions and that you should follow them. Of course, if the instructions are dangerous or highly questionable, a follower is permitted to inquire about them and, depending on just how dangerous or questionable they are, even not do them. For your essence, the most important instruction to follow is the Golden Rule.

Leadership qualities are more extensive. Kurt Lewin and his colleagues have identified four different kinds of leadership:

- **Dictator.** This is someone who uses fear and threats to get things done, and makes all the decisions. Only the most subdued follower will work effectively with a dictator.

Your Essence Traits

- **Authoritarian**. This person doesn't encourage any suggestions that don't match his/her own. The authoritarian doesn't trust anyone and operates much like a dictator, without obvious fear and threats. Often people considered "strong" managers are authoritarian and their style highlights the difference between management and leadership.
- **Participative**. This person favors decisions by group consensus. If the leadership isn't stifled by group paralysis, a participative leader can be the most effective because others see themselves as active stakeholders in running the activity. Such a feeling encourages everyone to give their best because things are being done the way they have decided.
- **Laissez Faire**. This person doesn't actually lead, just follows the direction of what other group members want. Think of a flock of birds following the lead bird when suddenly the flock shifts to a different leader and a different direction. Laissez faire leadership has the specter of chaos always hanging around. Even so, activities run as laissez faire are

better than those run by a dictator and often as good as those run by an authoritarian.

Keep in mind that, as is the case with most kinds of categorization, people don't fit nicely into a single bucket. Usually there's a mix of leadership qualities within an individual, with one exhibited at one time and another at another time. There will be an overall impression of the person, however, and that becomes the bucket in which they become labeled.

Improving Your Leadership

The United States Army uses the following eleven principles to generate effective leadership:

- **Know yourself and seek self-improvement** - In order to know yourself, you have to understand your *be, know,* and *do,* attributes. Seeking self-improvement means continually strengthening your attributes. This can be accomplished through self-study, formal classes, reflection, and interacting with others.

Your Essence Traits

- **Be technically proficient** - As a leader, you must know your job and have a solid familiarity with your employees' tasks.

- **Seek responsibility and take responsibility for your actions** - Search for ways to guide your organization to new heights. And when things go wrong, they always do sooner or later -- do not blame others. Analyze the situation, take corrective action, and move on to the next challenge.

- **Make sound and timely decisions** - Use good problem solving, decision-making, and planning tools.

- **Set the example** - Be a good role model for your employees. They must not only hear what they are expected to do, but also see. *We must become the change we want to see* - Mahatma Gandhi

- **Know your people and look out for their well-being** - Know human nature and the importance of sincerely caring for your workers.

- **Keep your workers informed** - Know how to communicate with not only them, but also seniors and other key people.

- **Develop a sense of responsibility in your workers** - Help to develop good character traits that will help them carry out their professional responsibilities.

- **Ensure that tasks are understood, supervised, and accomplished** - Communication is the key to this responsibility.

- **Train as a team** - Although many so called leaders call their organization, department, section, etc. a team; they are not really teams...they are just a group of people doing their jobs.

- **Use the full capabilities of your organization** - By developing a team spirit, you will be able to employ your organization, department, section, etc. to its fullest capabilities.

Socialization

The extent of how you fit in with a group. Range – **anti-social** to **monophobic**.

Evaluation:

Anti-social	= 0
Isolated	= 4
Socially accepted	= 10
Actively social	= 10
Monophobic	= 0

Your Essence Traits

Socialization is very culture-dependent, where the culture you live in dictates the forms of socialization that are acceptable and those that are not. Two species of chimpanzees are often used to illustrate differences in socialization. The common chimpanzee is known to be quite aggressive, with patrols of males looking for males from neighboring communities who are traveling alone, often killing those unfortunate travelers. The bonobo, on the other hand, is known for its peaceful nature and has been referred to the "make love, not war" species for its interest in sex over violence. Common chimpanzees and bonobos live in the same general environment, separated only by the Congo River in Africa, so their differences in socialization are of great interest to primatologists.

Human cultures also illustrate differences in socialization. For example, the Semai tribesmen in Malaysia are gentle people who don't like violent, aggressive individuals. In contrast, the Yanomamö Indians between Venezuela and Brazil try to be tough and aggressive. As with the ape example, these human cultures both live in environments with plenty of food and other resources, so their differences in socialization are a reflection of aspects of their culture rather than of their habitat.

Socialization is a learning process that starts when we're very young and continues throughout life. Different cultures do things very differently. What would you do if your baby wouldn't stop crying and it didn't have any physical needs (feeding, changing, or sickness)? If you're a Navaho Indian from North America, you'd remove the baby from the culture by taking it to a safe place and leaving it there until the crying stopped. If you're a typical United States mother you'd hold the baby and try to comfort it until the crying stopped. Navaho children learn that such behavior takes them away from people they love and, consequently, they tend to grow up to be quieter. United States children learn that such behavior brings them warmth and close companionship, encouraging more of the same. However, if a mother in the U.S. followed the Navaho procedure she'd be considered uncaring or even negligent. Cultural mores are a major part of the society in which we live.

The socialization you learn as a child and refine as you grow up is sometimes referred to as your *primary socialization*. You learn other kinds of socialization by

Your Essence Traits

understanding how the groups you associate with interact with the larger society and how you should socialize when moving to a different kind of career or a different level within your career. Without any dramatic personal events the kind of socialization we have as adults is a mixture of many other aspects of our personality (including warmth, emotional stability, privateness, etc.) and the experiences we have with our family, friends, religion, education, and the mass media. If your family expresses discriminatory opinions about others (race, religion, culture, etc.) then you tend to grow up with those same discriminatory attitudes. Parents who are abusive to their children usually had abusive parents themselves. The more alike your learned socialization aspects and experiences are the more ingrained your socialization level becomes, while the more complicated the mixture the easier your socialization level is to change.

War and personal attacks (physical or otherwise) are dramatic events that often inhibit your socialization because you think it's harder to know whom to trust; natural attacks (earthquakes, etc.) are dramatic events that tend to make people more socialized because it's easier to survive if everyone helps each other; terrible health issues in yourself or a loved one are dramatic events that can make some people more socialized while making others more inhibited.

Some dramatic personal events are chosen by the person rather than being thrust upon their life. Chosen events include moving to another culture, changing your religion, or changing your sex. These personal choices are often intense and require you to also change your socialization understanding, although you don't need to change your socialization evaluation level.

People with an **anti-social** level of socialization actually have a behavior disorder defined by the American Psychiatric Association as "...a pervasive pattern of disregard for, and violation of, the rights of others that begins in childhood or early adolescence and continues into adulthood." Sometimes such people are called *sociopaths* or *psychopaths*. If that's you, hopefully you understand why you get a zero for socialization. If you're not sure but have many of the following symptoms, you're anti-social. Although the symptoms sound awful, don't worry too much about it if you only have one or two of them – many people do.

- Persistent lying or stealing
- Superficial charm

Your Essence Traits

- Apparent lack of remorse or empathy; inability to care about hurting others
- Inability to keep jobs or stay in school
- Impulsivity and/or recklessness
- Lack of realistic, long-term goals — an inability or persistent failure to develop and execute long-term plans and goals
- Inability to make or keep friends, or maintain relationships such as marriage
- Poor behavioral controls — expressions of irritability, annoyance, impatience, threats, aggression, and verbal abuse; inadequate control of anger and temper
- Narcissism, elevated self-appraisal or a sense of extreme entitlement
- A persistent agitated or depressed feeling (<u>dysphoria</u>)
- A history of childhood conduct disorder
- Recurring difficulties with the law
- Tendency to violate the boundaries and rights of others
- Substance abuse
- Aggressive, often violent behavior; prone to getting involved in fights
- Inability to tolerate boredom
- Disregard for the safety of self or others
- Persistent attitude of irresponsibility and disregard for social rules, obligations, and norms
- Difficulties with authority figures

At the other end of the anxiety scale from the anti-social types are people with a **monophobic** level of socialization except that they, too, have a recognized behavior disorder. *Monophobia* is an acute fear of being alone. Does this sound like you? Then you do everything you can to have others close by all the time, sometimes even when you use the lavatory. Your hidden reasoning for behaving like this is your fear of a sudden panic attack that will cause you to do something terrible, collapse, or die. You may be indistinguishable from everyone else, perhaps even more outgoing than most, when accompanied by someone you trust.

Monophobia is different from the separation anxiety that children often experience when their parent (especially their mother) is not with them for a short period of time, although the feeling is the same. Monophobics, in fact, prefer their mother to anyone else, probably because of the feeling of safety she projected when they were children.

People with an **isolated** level of socialization are more victims than victimizers, although sometimes they cause their own distress. Isolated people are different from loners who are alone by choice and exhibit at least a mild level of anti-social behavior. Isolated people are alone because people don't want to be around them. Does this sound like you? It may be because you're somewhat anti-social and your behavior makes you undesirable as a companion. It may also be because you think most people are inferior to you and you only want to interact with those worthy of your time and attention. In that case, your isolation is good for everybody.

Isolation occurs more frequently in children than in adults because the socialization learning of children sometimes results in their forming groups that are nasty to people outside of their group. If the nastiness is directed towards a naturally shy child who has more difficulty making friends, that child may become isolated. This gives the group a feeling of power and (they think) the person affected is outside of their group and doesn't matter. Interestingly, this kind of group behavior persists as the children grow up, with teams, fraternities, and even academic groups like Honor Rolls. The result is that by adulthood almost everyone is part of some group and has that group cohesiveness, but that almost everyone feels isolated in some other context. It's only in the case where you really don't belong to a group and you feel excluded from activities going on around you that you have become an isolated person.

Socially accepted and **actively social** people have no socialization problems except, perhaps, having too little time to be alone. The difference between the two is that socially accepted people are often invited to be part of activities because they have such a pleasing personality and contribute nicely to the events, while socially active people seem to be *always* invited. Socially accepted people throw parties and host events; socially active people are *always* throwing parties and hosting events. Being socially accepted or socially active seems to be one of personal preference – if you want to make being part of activities a major focus in your life, do more of it and you'll be viewed as socially active.

Money is a relatively unimportant part of the social life of an individual, irrespective of what you see in the media. You can do things that don't cost anything and have as full a social life as the guy in the limo. The important personality traits for high quality socialization are that you enjoy being with people, that you enjoy having other people be with you, and you don't expect people to do things they don't want to do. In other words, smile and relax!

Improving Your Socialization

An individual with poor socialization, one who feels disconnected with others and possibly with the world in general, is a very unhappy situation for the essence. Such

Your Essence Traits

individuals may actively try to damage or disrupt others around them, especially those who appear weaker than the social misfit. If such individuals become leaders (which can happen depending on their other essence attributes, physical body attributes, or cultural conditions), significant wide-ranging damage may occur to physical bodies, cultures, and habitats beyond their own.

If you're an **anti-social** person, it might be a physical body brain disorder that's only helped with medication. If your brain is OK, you may have developed your anti-social behaviors as a child from experiences you had at home. There have been numerous studies linking a father's alcoholism or a mother's lack of affection for her children to their children's anti-social behavior. If you're really anti-social, seek professional help.

If you're an **isolated** person and think it's because you have some recognized anti-social behaviors, correcting them will make you a new person in the eyes of those you want to socialize with. If it's because you think you're too good for people (fat-headed), put your head on a diet. Nearly everyone is excellent in something and you should appreciate their excellence, and you're certainly not excellent in everything. That's why so many kinds of evaluations exist – for math ability, language ability, music ability, art ability, military ability, etc. There's the story about the express checkout line at the supermarket in Cambridge, Massachusetts, where the sign says you must have fewer than 15 items. If you see a student on this line with more than 15 items in his/her cart, s/he's either an MIT student who can't read or a Harvard student who can't count.

If you're a **monophobic** person you have an anxiety that's telling you, wrongly, that being alone is dangerous for you. You can't be talked or bullied out of your beliefs because your feelings are not based on an intellectual level. First you have to have others understand what "being alone" means to you. Some monophobics need to have someone in the room with them, others just need to know there are people somewhere in the building.

The way to start improving is to be alone for a very short (and known) period, gradually extending it as you hopefully become more comfortable. You might start by having your companion walk out of the room and walk back in, then maybe walk out of the building but be in sight before coming back in. You have to have a companion who understands that leaving you alone for two minutes doesn't mean leaving you alone for three. Of course this will take time, but monophobia is a personality disorder and you're trying to improve your personality.

However, if you're tired of being lonely but don't know what to do about it, here are some tips:

- **Learn to become more self-sufficient.** People sometimes feel lonely and helpless when they lose a spouse or their children have left home. This is also a time you're prone to victimization by thieves. To help you feel more self-sufficient, most communities have courses available to help you learn skills like insurance, auto repairs, and plumbing.
- **Develop a hobby.** If you already have a hobby, spend more time getting better at it. If you don't have one, think of what you like to do and go do it. You'll find there are lots of similar folks out there to do it with you.
- **Join a group.** This is a bit like developing a hobby, with your hobby being the group activities. There are groups for everything.

Your Essence, Your Eternity

- **Get a pet.** Pets are great companions. Of course, if you choose a cold-blooded lizard it might say something about the kind of person you are.
- **Help others.** There are plenty of people out there in worse shape than you are, and some are even lonelier. Sometimes all you have to do is listen.
- **Get out of the house.** You're not going to get less lonely if you stay at home alone. Surely there's some place you like to be. Go there.
- **Take a course.** This is different from the self-sufficient courses you may need. These courses are for fun subjects you never got around to. Always wanted to learn a foreign language? Want to out-trivial your friend Ken on the Civil War details he's so proud of knowing? Now's your chance.
- **Exercise.** Sitting around the house moping will also make you fat. Get out and run, walk, swim, or play something you like. Depending on what you choose you might have to endure jokes:

    > Doctor: This examination shows you're a complete physical wreck. You must avoid all exercise of any kind.

Patient: Does this mean ...?

Doctor: Yes. You have to take up golf.

Agreeableness
Compassion vs. Antagonism

Ethics

The extent to which you follow the right action for the greater good. Range: **full self-sacrifice** to **full self-gratification**.

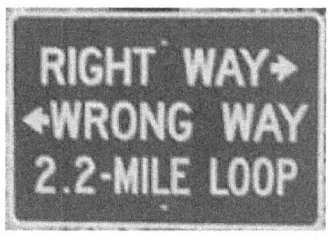

Evaluation:

Full self-sacrifice	= 3
Partial self-sacrifice	= 10
Neutral ethics	= 5
Partial self-gratification	= 3
Full self-gratification	= 0

Ethics and morals are often used and described as the same concept, but to understand personality we need to focus on them individually. Ethics is defined as "the principles of right and wrong" while morals (see next section) is following the rules of the society and the Golden Rule. Generally restricted to human conduct, our view of the essence broadens ethical behavior to include any living thing capable of it. There are also specific groups that have their own sets of ethics (such as medical ethics and business ethics) and those behaviors are included for individuals involved in those groups. Of course, even criminals have their codes of ethics, so gauging true ethical behavior can be more subjective than you think. If someone is a very ethical criminal, would you want him as your neighbor? Greed and power are often linked to poor ethical behavior. For our personality evaluation, an individual's total ethical behavior for the common good is measured. If a person only thinks about him/herself, they're **full self-gratification** and have no ethical values. The highest ethical values go to people who **partially self-sacrifice** themselves, because while those people respect the common good more than their own interests, they temper that with some recognition of their personal needs. After all, to **fully self-sacrifice** yourself for the common good is to have no respect for yourself.

Here's a list of common core ethical values from a variety of sources:

- Integrity (steadiness)
- Honesty (truthfulness, sincerity)
- Fidelity (trust)
- Charity (kindness, caring)
- Responsibility (reliability, dependability)

Keep in mind that an individual's ethical behavior can change over time. A person can be a no-good greedy bum and then later in life change to a person who does everything he can to make the world better. His ethical evaluation would be a reflection of when it was measured.

Improving Your Ethics.

There are two separate but related aspects of ethical improvement. The first, and most important, is improving your core ethical values. The second is improving your professional ethics. Each profession should have ethical guidelines to follow and adhering to them should satisfy the ethical needs of your profession. It goes without saying (hopefully) that if the ethical values of your profession are in conflict with the common core ethical values ("Don't cheat more than 10% of your customers") then your profession needs help.

We'll concentrate on improving your core ethical values:

- **Improving integrity** - From time to time important issues will come up and your opinion will be sought. When you see that an issue is important, educate yourself on the various aspects of it and decide which one is right for you. If you take a stand against immigration but hire foreign workers to clean your house because they're cheaper, you'll be seen as a hypocrite with little integrity. Politicians are viewed as having no integrity when they flip-flop on issues to satisfy a special interest group. Of course you can change your opinion on an issue, but your reasons should be for the common good rather than for personal gain.
- **Improving honesty** – Ever hear the expression, "Honesty is the best policy"? If you haven't, I'm sure your parole officer has. If you're someone who lies and cheats all the time, you're not honest. Of course, sometimes a little lie may be hard to avoid. For

example, if your wife asks "Does this dress make me look fat?" there's no answer that includes the word "yes" that will continue tranquility in the relationship. Honesty is sometimes associated with conflict, but if you avoid conflict by hiding your true likes and dislikes you'll be seen as insincere and your honesty in other (possibly more important) areas will be questioned. To maintain close relationships with the people you want in your life, focus on honest communication. Without it, people won't understand you as well and your ties with them will be weakened.

- **Improving fidelity** – Fidelity is being seen as trustworthy. Trust is something that's often hard to earn but easy to lose, so be careful how you handle positions of trust. You can lose a person's trust if you abuse them (physically or emotionally), express hostility towards them, see them as a competitor, or express such low self-esteem that your ability to maintain trust is questioned. Interestingly, you gain a person's trust by being seen as vulnerable because the other person feels more intimately involved with you and also senses a bit of control over you. Please note the difference between low self-esteem and vulnerability. You can also improve people's trust in you if you have trust/confidence in them. You can see how this works – if someone has confidence in you, you respect their good judgment and correspondingly have confidence in them.

- **Improving charity** – Being hard-hearted is neither charitable nor ethical. But being too soft-hearted permits you to be taken advantage of by people who are neither charitable nor ethical. The best approach is to be kind and caring towards those who are less fortunate, without making yourself unfortunate. Some people express their charitable nature by giving their money and/or time to recognized charities, and often it's that kind of support that keeps those essential charities operating. While commendable, that's not the kind of charitable motivation expected of an ethical person. This is: If you're on public transportation and you see someone having trouble maintaining their balance, offer them your seat; If your sister will be having a medical procedure done in a hospital for a couple of days and her husband is away, offer to take care of her children while she's gone; If you see a lost dog on the street, try to catch him and find his owner (which may require various inquiries over several days) or find a suitable new home for him. Being charitable is all about relationships between you and other living things associated with you, and the way you handle those relationships. If you're not a naturally kind and caring person (many of us aren't), you may need a new view of life to achieve the charitable nature you should have.

- **Improving responsibility** – Some responsibility situations are easy. It's always irresponsible to drive when you're drunk, even if you don't hurt anyone or any thing. It's always irresponsible to tell people you'll be somewhere or do something, and then purposely not be there or do it. When you get married your spouse expects you to be a responsible person. When you get hired by a company your boss expects you to be a responsible employee. Those are all normal expectations of life – if your spouse doesn't think you're responsible, they will let you know it through their obvious displeasure with you. This is expressed in the song *Marry the Man Today* from the musical *Guys and Dolls*, part of which is:

Your Essence, Your Eternity

> Marry the man today,
> Give him the girlish laughter.
> Give him your hand today
> And save the fist for after.

If your boss is unhappy with your performance, you'll certainly find out about it. Other situations of responsibility don't seem to be as firmly established. For example, your aged parents live nearby but you haven't contacted them in years. Must you do that, or did your responsibility to them end when you became a legal adult and moved away to your new life? And are you responsible for finding them the proper medical care when they can't do it themselves? How about your children? While parental responsibilities are prescribed by law, are you responsible for sending them to college? What if you have 10 kids? Are you responsible for helping a stranger who's being robbed? Many of these situations are not clear-cut. To be the most responsible person you can be, you should refer to the Golden Rule: "Do to others as you would have others do to you." If you follow that in each situation, you will be a responsible person.

Your Essence Traits

Morals

The extent to which you respect the principles of society. Range: fully self-indulgent to fully rule-bound.

Sign near an airport moving escalator

Evaluation:

Fully self-indulgent	=	0
Partially self-indulgent	=	3
Neutral morals	=	5
Partially rule-bound	=	10
Fully rule-bound	=	3

Are you a moral person? Few of us would say we have no morals because that implies we care nothing for anyone or anything other than ourselves, and that we can't be trusted in anything. You could be *amoral*, which generally means you don't feel subject to standard moral values and lack a feeling of responsibility for your actions. Surprisingly, some leaders are amoral but you'd probably prefer to be labeled as amoral than as *immoral*, which usually implies sexual misconduct. More leaders have lost their positions because of improper sexual behavior than for any other reason, and the shocked public loves every minute of it.

Besides proper sexual behavior, how you view God is also an indication of your morals, at least according to a Pew Research Center poll taken in 40 countries from 2011 to 2013. Across Africa and the Middle East, only in Israel does a majority think it's not necessary to believe in God to be a moral person. If you live in a poorer country belief in God is more important than if you live in a wealthy country, with two big exceptions. Americans say belief in God is essential to morality while the Chinese don't think so.

The *Stanford Encyclopedia of Philosophy* defines morality in one way as referring to a code of conduct put forward by a society, some other group (such as a religion), or accepted by an individual for his/her own behavior. For the purposes of your essence the morals of an individual are measured only by the code of conduct put forward by their society, plus their adherence to the Golden Rule as the universal baseline for moral behavior. Keep in mind that different societies have different morals and you can only judge the morals of a society within the context of that society, not in the context of your own. For example, Fons Trompenaars reported the following results when people from different cultures were given a moral situation. They were a passenger in a car driven by their friend who had been driving too fast and struck a pedestrian. Would they lie to protect their friend?

> **United States.** Passengers lied about the accident while in the presence of their friends, however, when isolated they were generally willing to 'drop hints' about the actions of their friends.
>
> **Russia.** After some coaxing, the passengers were willing to admit that their friends had committed the crime, and even admit to crimes that their friend the driver had not even committed (e.g. prostitution, drug smuggling, arms dealing).
>
> **Brazil.** The driver and the passenger were both inclined to create highly fictitious and unbelievable stories about how the driver was not at fault. Generally, they posed that the pedestrian seemed to be attempting to commit suicide and had, therefore, jumped in front of the car.
>
> **France.** After being plied with cigarettes, we were able to convince the French passenger to confess to the crime rather quickly.

Kenya. Kenyan drivers were apt to dismiss that they had committed the crime even when the evidence was shown that they were clearly responsible for the killing of the pedestrian.

Saudi Arabia. In most cases, the passenger of the car was female and claimed that she would have been unable to witness the incident due to the religious covering (*hijab*) which she had been wearing at the time.

Italy. The Italian passengers generally responded to police interrogation with 'hot-blooded' responses, often stating that the dead pedestrian had insulted the driver and that it was his natural recourse to run the pedestrian over.

In all cases, however, the Golden Rule rules. If the morals of an organization to which the individual belongs are contrary to the morals of their society or to those of the Golden Rule, and the individual adheres to the organization's morals, we consider that individual to be amoral. If an individual has a personal set of morals that are contrary to those of their society, that individual is amoral. Criminals who like their criminal activities have such morals.

Of course, some societies have been (are?) criminal, as the Nazi government was. In such a case, an individual with a personal set of morals contrary to those of their criminal society would certainly not be amoral. But you can see how such reasoning can become a trap – what's criminal to one person can be normal behavior to another. We always have to go back to first principles – if the Golden Rule is violated, the morals involved are not moral.

A key to having a moral society is to train our children in moral values. If you're like most people, you're concerned that today's youth has rejected much of the moral code their parents were brought up with, raising fears about the future leadership of our country and its institutions. Your concerns are not new. Consider the following quote, attributed to Plato:

> What is happening to our young people? They disrespect their elders, they disobey their parents. They ignore the law. They riot in the streets inflamed with wild notions. Their morals are decaying. What is to become of them?

Jean Piaget believed that children create their conceptions of morals based on their observations of the world. He said:

> The child is someone who constructs his own moral world view, who forms ideas about right and wrong, and fair and unfair, that are not the direct product of adult teaching and that are

often maintained in the face of adult wishes to the contrary.

If you think only of yourself you're **fully self-indulgent** and have no moral values. If you focus somewhat on yourself (**partially self-indulgent**) or follow the rules of your society completely and without thought (**fully rule-bound**) it's better. You have the highest moral values if you're **partially rule-bound**, following the moral rules of your society but recognizing weaknesses in them. In this case, you help your society grow with better understandings of the moral rules that are needed.

When the moral positions of a society are changed (marijuana is permitted/forbidden, abortions are permitted/forbidden, etc.), which one is "right"? Often that's not an easy question to answer, but we can answer it by referring to the Golden Rule: "Do to others as you would have others do to you."

Peterson & Seligman have found that certain attributes have been considered moral across societies for thousands of years. Those attributes are:

- **Wisdom** – correct judgment
- **Courage** – meeting opposition with firmness
- **Humanity** – being humane

- **Justice** – conformity to the law
- **Temperance** – being moderate
- **Transcendence** – understanding beyond experience

You could be basically moral but have an event or two that would be considered not moral. These events must be recognized and corrected so that, over time, you'll have fewer lapses. You can learn from others, picking the best of their morals to include in your own and recognizing the worst to avoid. While this sounds sensible and straightforward, actually succeeding at moral improvement requires dedicated attention. One approach, established and successful for improving the way things are done in industry, is referred to as *continuous process improvement*. A continuous process improvement model developed at the University of Maryland and used by the National Aeronautics and Space Administration (NASA) can be reworded and used for your personal moral improvement. See Chapter 12.

If each individual in a society followed continuous process improvement methods to improve their morals, eventually the entire society's morals would be improved.

Your Essence, Your Eternity

Don't think it can happen? Individuals banding together have historically been the driving force for changes in civilization at all levels. In 1967, during the Vietnam War protests, folk singer Arlo Guthrie wrote the song *Alice's Restaurant* that included the following lyrics to encourage such individual actions:

> You know, if one person, just one person does it they may think he's really sick and they won't take him. And if two people, two people do it, in harmony, they may think they're both faggots and they won't take either of them. And three people do it, three, can you imagine, three people walking in singin' a bar of Alice's Restaurant and walking out. They may think it's an organization. And can you, can you imagine fifty people a day, I said fifty people a day walking in singin' a bar of Alice's Restaurant and walking out. And, friends, they may think it's a movement.

Improving Your Morals.

Because morals vary across cultures, moral improvement is restricted to improving your universal moral attributes noted above and further described as follows:

- **Improving wisdom** – First, the bad news. You don't necessarily get wiser as you get older. The good news is that you don't have to grow a beard and live in a cave on a mountain to be wise.

While the most important factors in attaining wisdom are through knowledge and experience, you have to know how to use them. Lacking specific knowledge and experience, you can also attain wisdom through logical thought and reflection (the foundations of Aristotle's philosophy and in teachings of all the major religions), but those who follow the *freethinking* philosophy insist that you have to have knowledge before making any judgments.

There are various lists of attributes of wise people, but in general a wise person recognizes the most important parts of a problem rather than just seeing the problem, and a wise person's opinion is consistent with their core ethical values. According to Confucius, the easiest way to attain wisdom is to find someone you think is wise and imitate them.

- **Improving courage** – Courage can be expressed in various ways. For example, you might see people trapped in a burning building and rush in to save them. You might be terrified of public speaking but give a stirring oration at a friend's funeral. You might have to admit to your parents that you're in financial trouble and need their help. All of these events take courage. As Audie Murphy, the most decorated U.S. soldier in World War II said, "I'll tell you what bravery really is. Bravery is just a determination to do a job that you know has to be done." The events just mentioned all apply as moral courage in additional to physical courage because no one acts courageously if they aren't convinced it's the right thing to do. Here are some suggestions for becoming more courageous from the Metcare Health Newsletter of August 2009 and from Dr. Paul Stoltz, author of *Adversity*:

- Recall previous times when you acted courageously. Applaud yourself for showing such courage.
- Shift your focus. Don't worry about what others think, only what you believe to be the right thing to do.
- Eliminate the words *wish, hope,* and *maybe* from your vocabulary. Those words erode courage by creating doubt and hesitation.
- Do your homework. Know the obstacles you may encounter, but understand that all the obstacles may not be known until they occur. Don't worry about problems that don't exist until they actually exist.
- Avoid naysayers. There are always people who will say something can't be done, but if you

believe in yourself and your goals, you should try the best you can. You don't want to be someone who says, "I could have done it if I really wanted to."

- **Improving humanity** – The way this attribute is worded may confuse its real intent. The word "humanity" and its description of "being humane" imply actions only for humans, but the intent is moral behavior for all living things. This becomes clearer when you consider the U.S. organization called the *Humane Society*, which is really about animals and not about humans at all. So, when you review your own humanity, do it in a more comprehensive way to include your moral perspectives for all living things.

    In general, then, your humanity includes your moral attitudes towards health, food, and the environment as they apply to all living things. If you see an injured or hungry bug you may not feel compelled to help it. However, if you see a family pet with these problems, you should do what you can to help. In these cases, it's really a difference between "wild" and "domestic" animals. Wild animals are expected by nature to look after themselves, which means that "wild" includes you. Domestic animals, however, are under your care, and your morals should recognize that responsibility. Naturally, the same thing holds for plants.

Your moral attitude towards the environment is more straightforward – if you intentionally damage the environment, you're immoral. To improve your humanity towards the environment, find some way to make your environment better. This could be as simple as picking up trash or planting flowers. It's all part of the Golden Rule.

Improving humanity focusing on humans includes areas such as agriculture, the arts, social development, culture, education, and religion. Do you think it's moral or immoral to grow tobacco? At what point do works of art offend your moral sensibilities? If your society permits capital punishment, do you think it's moral or immoral? The rules in some cultures may be different than your own. For example, is it moral or immoral to have multiple wives? Is sex education moral or immoral? Do you think any religion is immoral? Is working for peace always the moral thing to do? As you can see, many of the human-oriented moral questions have become social questions, often decided by legislation or decree. The culture in which you live dictates the framework you can use to address those questions, and your success at improving humanity will be partially based on how well you do that.

- **Improving justice** – Moral people obey the law; immoral people disobey the law. What could be clearer than that? But wait – have you cheated on a test, purposely driven faster than the speed limit, or copied a song or movie without paying for it? There are laws against those actions, too, even if "everyone does it".

  Before we look into improving our justice morals, let's consider why people obey the law at all. George Klosko suggests that laws are written to promote three principles: fair play, natural duty, and the common good. But people have to believe that a law is fair, that they have an obligation to obey it, and that it's really the best way for everyone. And that's why you copied that movie – you aren't convinced that law is one you must obey. Also, obligation to obey the law decreases considerably if people believe the legislators who make the laws don't obey them.

There are certainly some laws that must be obeyed by everyone, no matter what, because the situations they address are considered major crimes. The most significant crimes noted by the FBI are criminal homicide, forcible rape, robbery, aggravated assault, burglary, larceny-theft, motor vehicle theft, and arson. If you do any of those you're immoral, period. To improve your justice morals in those cases, accept the penalty you receive for committing the crime and don't do it again.

Improving your justice morals when you disobey the law in other cases is not as straightforward. For example, let's say you're driving on a main highway at the posted speed limit when you round a curve and see a new (much lower) posted speed limit and a policeman with a speed radar gun. You're given a ticket and must pay a fine. Are you immoral for breaking the law or is the government immoral for running a speed trap? If you decide that justice requires you to fight this ticket, then you should go to court and try to improve governmental morality. Another example – you're doing your income tax and heard that the IRS only inspects charitable deductions if you claim more than $1000. You only gave $20 to charity last year, but you put $900 on the income tax form. Are you immoral, even though you're within the IRS guidelines? Yes. You're also a cheapskate. To improve your justice morals, reflect on the principles of fair play, natural duty, and the common good. In all cases, of course, if the law in question violates the Golden Rule, you should help to change it.

- **Improving temperance** – Moderation is a classic virtue that's part of many cultures. It's one of the four cardinal virtues in ancient Greek society and a virtue in many contemporary religions. Temperance generally involves having moderation in the great pleasures of eating, drinking, vanity, and sexual interests. The thinking is that if you can't control your lusts, you can't be a moral person.

  Since moderation is linked to so many religions, adherents have been emboldened to start temperance movements, initially to ban alcoholic beverages. Successful movements were established in the United States, Britain, Ireland, New Zealand, Canada, and Sri Lanka. In the U.S. they convinced Congress in 1919 to pass the 18th Amendment to the Constitution, banning the sale, manufacturing, and transportation of alcohol for consumption, and the result of this Amendment was Prohibition. The outcome was less than satisfactory, with illegal bars (called *speakeasies*) opening up everywhere. By 1925 it was estimated there were between 30,000 and 100,000 speakeasies in New York City alone. When the Great Depression hit in 1929 Prohibition became increasingly unpopular, and it didn't help that President Herbert Hoover called it "The Noble Experiment". Hoover lost the election in 1932 and Prohibition was repealed shortly afterwards. The English writer G. K. Chesterton called Prohibition "an intemperate denunciation of temperate drinking".

While it might seem that people who advocate temperance are cheerless dried-out prunes who want to take away all your fun, that's not the case (usually). Temperance is avoidance of excess. By following a temperate life you recognize what's needed to fulfill your desires and strive to attain those things, rather than trying to attain everything. For example, many overweight people spend much of their time and money on diets that don't work because they aren't temperate in their eating habits. Every drunk driver is intemperate. So many people are unhappy even when it seems they have everything they need because they want more. Kids, who are still immature, want toys (take a kid to a toy store and find out how much they think they need). There's a beauty of innocence in children, but there's nothing beautiful about a spoiled child, especially when that child is an adult.

Temperance is also involved in how you project yourself to others, your sportsmanship (both winning and losing), and your quickness to forgive someone's errors. Who would you rather hire as your sales manager – the man with large tattoos and a nose ring or the man without them? Sure, you should be able to dress and look the way you want, but you also have to understand how others will view your appearance. Which would you rather see – a losing athlete who says the reason she lost was that she made too many errors, or one who congratulates her opponent for playing better than she did? How soon do you forgive someone for doing something wrong? In some extreme cases, even the most temperate person will never forgive.

So, the way to improve your temperance is to look at each event or activity and set your personal restrictions on it. Be sure your restrictions are reasonable and moral. "If you don't stop screaming, you'll have to go to your room" is a reasonable and moral restriction for a child, not, "If you don't stop screaming I'm going to cut out your vocal chords." For more continuous lifestyle activities, set a long range "mission" with short-term goals that you can achieve. "I'll play tennis for four hours every week" may be reasonable but, "I'll play tennis for four hours every day" may put you in the hospital. If you think temperance restricts your freedom too much you might decide, as my friend Donald Graves remarked, "I'll be temperate in everything, including temperance. So, once in a while I'll have a really wild time."

- **Improving transcendence** – Transcendence, in the philosophical sense, originally meant the separation of God from the world rather than God being involved with the world, as an explanation for the impossibility of ordinary people to understand why God does things. This meaning has been expanded to include knowledge beyond any knowledge of a human being, the mysterious sense-of-self through which we experience the object world. Some alternative life styles, including *Transcendental Meditation*, are based on attempting to achieve this higher level of understanding.

For you, transcendence is really your essence. Transcendence includes understanding and the personality attributes associated with it, but not consciousness, which is a separate biological function that's part of the physical body. Transcendence may include much more, but we as humans in a human physical body are restricted to the transcendent capabilities permitted to the essence by that physical body. Of course, such restrictions hardly restrict our desire to know, and that desire is the driving force for improving our transcendence. How do we do it?

There are two aspects to consider. First, our basic physical body needs must be met before we can consider improving our transcendence. Second, we must keep in mind that superior moral understanding must go hand-in-hand with superior human knowledge so that we don't become an immoral super-smart person. Having both of those aspects firmly in mind, we can look at improving our transcendence.

From an essence perspective, there are three kinds of transcendence: a spiritual one, a humanist one, and a topical one.

Spiritual transcendence refers to the ethereal understanding of the essence of living things. These are the capabilities attributed to the founders of religions (Abraham, Buddha, Christ, Mohammed, etc.) and by individuals recognized as mystics. It's also claimed by people who have taken certain mind-altering drugs, and those claims have generated very vocal opponents and proponents. The opponents say the mind-altering drugs induce chemical changes in the brain that simulate spiritual experiences, demonstrating that all spiritual experiences are just chemical imbalances within the brain. "Blasphemy!" say the proponents, who suggest the mind-altering drugs stretch the brain to provide a pathway to the spirit world.

The complicated interaction of biochemical processes within the physical body part of the brain coupled with the professed transcendent understandings of these people is why this level of transcendence is often dismissed as mental ramblings. How do we know what's merely biochemical and what's actually transcendent? In addition, there have always been immoral people claiming to have transcendent powers but who really only want the power to take your money. Judging from the various (but somewhat similar) descriptions of spirituality from transcendent individuals, such information must be both hard to understand and hard to communicate. Spiritual transcendence is not something that can be learned or improved (if you don't have that special kind of brain, you don't have and can't have spiritual transcendence), but those of us without it can try to understand it as best we can.

Humanist transcendence is demonstrated by individuals who seem to have a more than extraordinary understanding of how humans should act. Recent examples are Mahatma Gandhi, Martin Luther King Jr., and Albert Schweitzer. You probably know of others. If you're not one of these rare individuals, your humanist transcendence will be at a lesser stage. You can improve your humanist transcendence by learning and emulating positive personality characteristics. If, instead, you choose to learn and emulate negative personality characteristics, you can look forward to a rather unhappy life, perhaps prison (or worse). Remember the Golden Rule.

Topical transcendence refers to individuals who understand specific concepts so much better than everyone else that those people are considered giants in their fields. While topical transcendence is really an aspect of the physical body, attaining it affects the essence. Albert Einstein is the classic example of a person with topical transcendence, but there are many others, in many areas. Great artists and musicians (for example) have as high a topical transcendence as great physicists. These people are sometimes referred to as *savants*, who were discussed more fully in Chapter 9 under *Essence Age*. Some achievers of topical transcendence are born with these abilities, some have gotten them after an illness or injury to their brain, and some have achieved their brilliance through personal effort. Wealth has nothing to do with it - the great mathematical genius Srinivasa Ramanujan Iyengar from India was born into such poverty that he had to use chalk and slate because he couldn't afford paper.

The way to improve your topical transcendence is the old fashioned way – practice, practice, practice. Pick something you're naturally good at and keep improving your abilities with it. After a while, perhaps a long while, you'll achieve such a level of capability in your topical area that you'll be considered a genius in it. Sometimes topical transcendence requires you to start by a certain age. But if you don't do the initial work or don't continue your practice, you won't achieve topical transcendence. Note 3

Sensitivity

Whether you're tough-minded or tender-minded. Range – **impersonal** to **extremely sympathetic**.

Your Essence, Your Eternity

Evaluation:

Impersonal	=	0
Recognizing weakness	=	2
Sensitive	=	10
Sympathetic	=	8
Extremely sympathetic	=	1

Individuals are born with an innate sensitivity from their essence. When you watch a movie containing graphic violence, do those images disturb you and remain in your mind long after the movie ended? Then your essence orients you towards a prey physical body. On the other hand, if you have feelings of excitement when watching a movie with graphic violence, your essence orients you towards a predator physical body. Anyone who has feelings of desire when watching graphic violence has a psychological problem and should seek professional help.

Your Essence Traits

While you're born with an essence, environmental factors immediately influence it. Are both of your parents psychiatrists? Then an elevated level of sensitivity is part of your upbringing. Are both of your parents professional wrestlers? Then you probably have less sensitivity within your home.

As you grow up, your sensitivity continues to be influenced by the people around you. If your friends bully others, your sensitivity is likely to be reduced; if they're the ones being bullied, your sensitivity is likely to be increased. But your friends are your choice, which probably means you're similar to them. If you've chosen your friends because they're more popular and you want to be more popular yourself, but you're uncomfortable with the way they act, choose different friends. It's not the right group for you.

If you have an **impersonal** approach towards sensitivity it means you act in a cool, even cold, manner to how events affect others. You think less of people when making decisions and more on the logical reasons for the decision. You do better in roles that involve ideas, data, or things, rather than in roles that involve people. That's fine for some aspects of life but much of what occurs happens because people with different, but valid, perspectives try to have their view become the accepted view. The term *fuzzy logic* from mathematics applies here, with data that's not so distinct. It's harder to be logical when logic isn't part of the arguments.

You don't necessarily view yourself as hardheaded, preferring to be considered more sensible than the softheaded people you see all around. Extreme individuals seem to be insensitive and even to lack emotions. Youngsters who suffer traumatic events in their lives sometimes withdraw from others and become impersonal in their relationships. Impersonal people usually lead more lonely lives because others shy away from people who act like they'd rather be alone. Sometimes, though, you'd really like companionship but you've been hurt by close relationships in the past and find the whole process of making friends to be difficult and threatening. You don't have to worry about keeping friends happy if you don't have friends. Because of the circumstances faced in fighting, you'd be a good soldier and military instruction includes training to become more impersonal. Lawyers try to avoid having impersonal people on the jury because they're unlikely to be swayed by arguments with logic holes and, worse, they'll point out those holes to the other members of the jury.

If your sensitivity level **recognizes weakness** you have some sensitivity to the effects of actions on others, but see those effects more as demonstrating weaknesses in other people than as situations affecting them. You wouldn't try to exploit those weaknesses (well, maybe you would), but seeing the world this way gives you a sense of superiority because you (apparently) don't have such weaknesses. That sense of superiority is your biggest problem – the more superior you appear the fewer relationships you'll have. No one wants to be with a fathead. You're rather tough-minded because of your view of other's weaknesses and of so few weaknesses of your own. Keep in mind that the weaknesses we're talking about only refer to views – if you're physically less attractive or less skillful in some activity you understand those weaknesses in yourself. In that way, you could have a sympathetic level of sensitivity for those weaknesses. Discussion topics are different, though. An interesting thing is that you can be on any side of an argument and confidently see the weaknesses in the other sides. You'll make a good politician or a good salesperson.

Your Essence, Your Eternity

Sympathetic people are on the tender-minded side, interested in reaching out to give their support to others with problems or issues. Does this sound like you? If the school needs volunteers to help tutor the kids, you'll volunteer. If a disaster relief fund needs money, say no more – a check is in the mail. When the Red Cross comes to town for a blood drive, you're there with your sleeves rolled up. You're a wonderful person with only one negative characteristic – you have a hard time turning down anyone in need. The result is that you're completely overbooked, with little time of your own. If you really can't turn down any good cause who asks, you may wind up working so hard that you develop physical problems.

Sympathetic people may not be the best type to have around because they may become an enabler for bad behavior. After all, if someone is acting badly and the people around them are sympathetic, why should they change? Some people who panhandle are doing it not because they're desperate, but as a career choice. They've found that many people are sympathetic to their "plight" and give them money to help, more money than the panhandler made working on their last job. You're not stupid, you just don't know if the panhandler is real or not and sympathetically decide to donate just in case.

You also try to sympathize with all sides of an issue, even when you believe in only one of the sides. You're one of the independent voters that are so actively sought in every election, where each side tries to convince you to sympathize with their positions. Lawyers with weak cases like sympathetic people on the jury because the lawyer (who has a streak of recognizing weaknesses in their character) feels they can play on the sympathies rather than rely on the evidence. If you're a sympathetic person, you have my sympathy.

The most desirable level of sensitivity is when you're sensitive to the perspectives of others while choosing a small selection of causes to help further. You have to be both tough-minded and tender-minded at the same time so that you can adequately support the causes in which you firmly believe while continuing to be sensitive to other causes. It's not easy to say "no" to a good cause, but sometimes it has to be done. Your tough-minded nature also helps weed out fallacious arguments and correspondingly fallacious causes. Interestingly, while rich people donate a lot of money to their causes and often get the attention of the media because of it, poor people are just as likely to donate a proportionate amount of their resources. Being rich is nice, but true sensitivity is attained through other means.

You might wonder why extremely sympathetic people get such a low evaluation score. Some of the reason is that extremely sympathetic people drive others crazy with their sympathy. Sound like you? Most people are only looking for a little recognition of their troubles, not the overwhelming assistance you offer them. Recipients often have concurring, but contradictory, feelings about the attention they're being offered. In one way, the recipient doesn't want to be seen as so helpless and hopeless that they need such support. In another way, the recipient may wonder if you have some hidden agenda and the recipient is somehow being set up. In yet another way, the recipient may see all this attention as favors that must be repaid in some way someday, and they may not want to be under such a debt. Overall, being helped by an extremely sympathetic person isn't as good as it might seem to be to someone from the outside.

Of course, there's also a subset of people who will take advantage, and even exploit, your behavior. This is different from the "cult" situation, where a person is required to give all their money to a group in exchange for acceptance. You donate as their way of helping people you see as less fortunate. While it's desirable to be able to sympathize with another's problems, going too far may adversely affect your objectivity. For example, if your friend always leaves work early but is fired, his reason that his boss "felt threatened because my work was always better than hers" might elicit a "that's really too bad" from you. However, if you genuinely believe your friend's explanation, you should re-examine how you evaluate sensitive situations.

There's another kind of extremely sympathetic person – the kind who is totally passionate about a cause and is equally passionate against those who have opposite views. You see these people frequently in the news, almost always because they've committed some anti-social act to support their cause. Extremely sympathetic people of this kind are the radicals of every organization and will be your trusted friend if you support their cause and your dedicated enemy if you're against it.

Improving Your Sensitivity

If your current sensitivity evaluation is **recognizing weakness**, you can improve by being able to understand situations and recognize the sensitivity to them by adherents on both sides of the arguments. This doesn't stop you from taking one of the sides, even adamantly, but you first have to be able to see both of the sides. One way to do this is to join a debating club. These clubs often have their members argue one side of an issue, then argue the other side. Getting practice in understanding and trying to convince others to believe a position you previously didn't believe in will broaden your outlook on other perspectives and, hopefully, help you see fewer weaknesses and more sensitivity.

If you're starting out with a more extreme sensitivity level you have more adjustments to do. If you're currently **impersonal**, do you really like being thought of as cold-blooded? If you do, you deserve your zero in sensitivity. However, if you want to improve your sensitivity level, start by sitting in a public place and watching people as they go by. No, this isn't an exercise for you to ogle physically attractive people. Try to notice their expressions and body language, especially people in groups, and imagine what they're thinking. This simple exercise will begin to connect you with situations in other people's lives. When you think you understand that other people's problems are important to them and shouldn't be ignored by you, take the next step by asking your friends to talk with you about what they've been doing recently. Ask them questions and be genuinely interested in what they're saying. The more you do this, the more attuned you'll become to other people and the more sensitive your personality will become. Impersonal people may have trouble doing much of this, but that's because you're normally impersonal. The idea is not to change your personality to be someone you're not but to broaden your sensitivity so that your interactions with others of your species will be more congenial.

If you're currently **extremely sympathetic**, you have the opposite situation. You may say you're just sympathetic to an issue, but you've gone far beyond that. Social issues are lightning rods for extremely sympathetic people on both sides and it's not uncommon for violence to be part of it. If you're involved in issues to this extent, you need to change your sensitivity. Here's what you do.

First, recognize that you've gone too far with the issue. You should set limits that must not be against your society's laws (as previously mentioned, if the laws in question violate the Golden Rule you should help work to change them). You may also have to remove yourself from the activist group if they've gone too far. Within this new framework, express your sympathies as you see fit. Not only will you have a far greater chance of achieving real and lasting success with your activism, you'll stay within the rules of your society and have more freedom to be a sympathetic activist with other issues.

Aggressiveness

The manner of your interaction with others. Range – predatory to loving.

Evaluation:
- Predatory = -5
- Tactless = 0
- Passive = 5
- Friendly = 10
- Loving = 2

Your Essence Traits

Describing your interaction with others is one of the most individually specific personality attributes you have because it depends a great deal on who the interaction is with. For the same situation your interaction with a loved family member will likely be very different from your interaction with a stranger. For the purposes of this discussion we'll assume that a friend has just introduced you to someone you've never met. The evaluation levels above are shown for this kind of interaction. If you're interacting with people you know, adjust the evaluation levels accordingly. After all, you'd better be more than passive with your spouse or you may be looking to be introduced to strangers!

We should set some parameters in our discussion of aggressiveness. First of all, we're talking about aggressiveness within a species, not between species of living things. A lion is a predator but its actions are based on physical body needs to ensure its survival, not the personality traits we mean to emphasize here. Aggressiveness can take several forms, including physical, mental, and verbal actions. We should also recognize that "aggressiveness" is not the same as "assertiveness", with an assertive person being less forceful than an aggressive one.

Finally, the type of aggression we'll discuss is the personal interaction kind, not the goal-oriented kind. Your sweet, loving girlfriend may seem quite different on the tennis court when she's going for set point against you.

Within a species, aggressiveness is often used to establish a dominance hierarchy, with the more aggressive members usually becoming the leaders. Humans have modified this a bit to conform to cultural expectations – the biggest bully on the block will more likely end up in jail than as a leader in society. The level of aggressiveness between people also depends quite a bit on the sex of the people – male-male, male-female, and female-female encounters should be separately considered. Aggressiveness has been linked to higher levels of testosterone, although suggesting that an aggressive female has more testosterone than her less aggressive male partner will likely get you a punch in the nose from both of them.

With all of those caveats in mind, let's look at the different levels of aggressive behavior. A person with a **predatory** level of aggressiveness is not a nice person and their negative evaluation score reflects that. Professional criminals have this trait, but anyone who searches out others who are weaker than they are to exploit their weaknesses acts like a predator. Obvious examples are the bullies on the block but the mental and verbal aspects of predatory behavior can mask the predator. For example, a predator may gain the trust of others with the intention of stealing their money through a complicated financial investment scheme. The old saying, "If it sounds too good to be true, it usually is" should be employed whenever you're offered such opportunities. Or a predator adult may verbally abuse children who feel helpless to confront the person for fear of being harmed in some way. People who abuse animals just because they feel they have that power are predators. If you see someone with this trait, call the police.

When looking at **tactless** people we also have to include the other people in the activity. What's tactless to one group is high humor to another, so if you overhear something from another group of people that sounds awful to you, go somewhere to avoid it. An exception to this is if the tactless person is publicly expressing him/herself rather than intending to restrict the actions to their group. In that case a more direct confrontation is acceptable, but escalating a small problem into a big problem is an unacceptable level of aggressiveness from you. Asking for assistance from those in control of the area is better. For example, if you're in a family restaurant and someone is loudly using expressions that aren't suitable for children, ask the restaurant manager to intervene. This approach has the tactless person in conflict with management, not just with another patron, and has a much better chance of being successful.

If you're tactless you never learned how to interact with others or don't care about the consequences when you do. Another possibility is that a conversation track has gotten away from you and you can't seem to stop it. This is sometimes referred to as "having your mouth in gear and your brain in neutral." If this continues the immediate situation will get worse, you'll be blissfully unaware you've caused an incident, and you'll likely cause another one in the future. Of course, if you're purposely trying to be rude there's no need for others to consider your feelings while correcting the situation. However, if you didn't really mean to be tactless, try your best to get out of it as quietly as possible.

If you have a **loving** level of aggressiveness you should restrict it to your loved ones, not to someone a friend has just introduced you to. In that case, your loving is awkward at best and progresses quickly to the weird. People don't like too much physical closeness from strangers and feel uncomfortable when their personal space is invaded. That's why most people feel best when other people are a certain distance away from them during conversations. Your loving aggressiveness violates that rule. You also tend to ask and give personal information that's inappropriate for a relationship at this level. It's not that you're trying to be invasive – you're just trying to be friendly – but your efforts are too friendly. What you also don't realize is that a loving level of aggressiveness is a career-killer because people will worry that accommodating this kind of interaction may allow others to accuse them of sexual misconduct. Surprisingly, you usually don't understand this and continue on your merry way, unaware that other people are trying to figure out a way to escape without being rude themselves.

If you have a **passive** level of aggressiveness, you aren't one of the leaders of society. You never take charge of a situation nor are you the first one to suggest a new approach. You aren't necessarily a dullard and aren't necessarily a non-contributor. You just don't like taking the risk that's involved in being in the lead. Of course, dullards and non-contributors are passive, too, so you have to recognize why you're being passive.

Your Essence, Your Eternity

There are times when even dominant people should be passive. Say you're attending the reading of a new play and the room is filled with professional playwrights. You've never been to one of these before. At the end of the reading the author asks the audience for their suggestions on improving the play. This is a good time to be passive. Any time you feel out of your depth is the right time to be quiet and let those experienced individuals take the spotlight. The reason is not so much that you'll sound like an amateur idiot, but really to give yourself the opportunity to learn something new.

If you have a **friendly** level of aggressiveness you have a lot of friends. You try to understand how others in the group perform and volunteer your services as needed. If another person is the designated leader, you follow their lead; if the group needs a leader, you're available. Possibly the best trait you have is your collaborative skill, being a team player in whatever position is needed.

Improving Your Aggressiveness

You may realize you have a **predatory** level of aggressiveness because you look for people to push aside and you have no close friends. Predatory aggressiveness is

Your Essence Traits

considered a major personality defect and if you don't change it you'll have social (and possibly legal) problems throughout your life. Fortunately, dramatic improvements can be had by simply following a rule often given to children – don't speak unless you're spoken to. No matter how much you think the world will improve with your essential participation, don't do anything unless someone asks you to. Ask for their advice or direction when you start, and ask before you do any more. If no one asks you to do anything, sit quietly. Given your history of predatory behavior, taking this improvement initiative will mean you'll initially be doing a lot of quiet sitting, so bring something like a book to read or a game to do. Being in the background is very hard for a predatory person but it's something you have to learn to do.

If it's been suggested that you're **tactless**, obviously by another tactless person, there are steps you can take to correct this inferior level of aggressiveness. First of all, recognize when you're saying or doing something that's not benign, making your actions possibly tactless. Then assume whomever you're with is feeling vulnerable or is especially sensitive to negative remarks. That won't always be the case, of course, but it'll help you see how your actions may affect such a person. If you think you might have said or done something tactless (the reaction of others is a clue), apologize quickly and put yourself in "safe mode" (as they say at NASA) by shutting down all but your essential functions.

If you're always bubbly around people you don't know, you might have a **loving** level of aggressiveness. Assuming that you're not trying to use sex in an inappropriate way (which would make you a predatory person), you're really just trying to be friendly but you're using the wrong methods. Step back a little with a polite smile rather than a laugh or a giggle, volunteer only when someone asks for a volunteer, and don't get too close to people at the bar after work. Note that you still do the things you've always done, but now they'll be in a friendly rather than a loving manner.

If you think you're too **passive**, assert yourself a little. Try to copy people you think are friendly, but take it easy because those with a genuinely **friendly** level of aggressiveness have gotten that way through practice and you don't have that experience yet. Start slowly by volunteering for activities you feel comfortable in doing, start making suggestions in areas you know, and start applying for tasks with more responsibility. If you're also passive with people, ask about joining them at lunch or for after-work activities. When you're with them, though, you have to be more than your usual passive self or they won't want you to keep joining them. Friendliness is the key here. By stretching your horizons and becoming less passive you'll find yourself with a more interesting and fulfilling life.

The level of aggressiveness for children deserves special attention. Children just entering school need to have or quickly develop the proper level of aggressiveness to succeed in both their academic and social developments. For example, they should be comfortable in asking questions, initiating conversations, and taking direction from adults. Sometimes children, especially physically larger kids or those from violent households, use aggression to bully other children. Naturally this is inappropriate and must be corrected for the good of the child and everyone around them. Some level of aggressiveness is good for all kids, though, because it helps guide them in conflict resolution. Here's some advice:

American Academy of Pediatrics: Set firm, consistent limits to help children self-monitor emotions and behavior; make sure all caretakers agree to the same limits. Provide examples of effective and socially acceptable ways of managing anger; be careful not to reinforce aggression with aggressive forms of punishment. Also, model acceptable behavior as a caretaker by managing your own temper. Remember that occasional outbursts are normal. If aggressive behavior continues for more than a few weeks, consult a pediatrician or mental health professional.

National Association of School Psychologists: Overly aggressive behavior can signify a social skills deficit; direct instruction, modeling, and coaching can help children acquire the skill of assertion, which as a replacement behavior may help prevent aggressive behavior. (*NASP Best Practices in School Psychology 2002*)

Nobody said bringing up kids is easy, and establishing their level of aggressiveness is one of the most challenging parts.

Fairness

The extent of recognizing other points of view. Range – **consciously biased** to **consciously impartial**.

Your Essence Traits

Evaluation:

    Consciously biased   = 0
    Superficially biased   = 2
    Both biased and impartial = 5
    Superficially impartial   = 2
    Consciously impartial   = 10

Usually, "fairness" is a broad character attribute that includes taking turns, playing by the rules, not taking advantage of others, etc. We're using a restricted definition, taking only the aspect of fairness that looks at recognizing other points of view. Fairness doesn't mean every point of view is equal, but it does mean treating everyone's point of view with respect. After all, that's the way you want them to treat what you say and think, so you should "Do to others as you would have others do to you."

In this case, as with many of the personality attributes, it's your overall fairness with everyone that's evaluated, not just your treatment of a single individual. In each of the evaluation levels, think about how you act in general, not just in special circumstances. For example, if you don't know anything about a subject or the person discussing it, you're more likely to be impartial; if you're an expert on the subject or you know the person quite well, you're more likely to be biased. Don't forget, being biased doesn't necessarily mean you're against what's being said; it means you've previously taken a position (for or against) and your position probably won't change.

Philosophers disagree about whether or not bias is acceptable, or even admirable, in some cases. *Partialists* believe that loyalty to spouse, family, and/or country may be admirable. *Impartialists* disagree, saying everyone should be treated equally. Another perspective is that impartiality is only necessary for those acting as a judge, umpire, or public official. When you evaluate your own fairness level, choose a single point of view for all of your decisions and decide on your general fairness level from that.

Dolores Albarracin and her colleagues found some indicators that make people biased. In general, people are guided by two different motivations, a desire to feel validated and a desire to know the truth. You shouldn't be surprised that people are usually biased when they're talking about politics, religion, or their ethical values. When looking for something to read, people are about twice as likely to choose information that supports their own point of view rather than an opposing point of view and people with closed minds are even more reluctant to read something different. Surprisingly, people who have little confidence in their own beliefs are less likely to expose themselves to contrary views than people who are very confident in their own ideas. Albarracin said, "For the most part it seems that people tend to stay with their own beliefs and attitudes because changing those might prevent them from living the lives they're living."

Interestingly, no matter how bone-headed wrong you think your elected officials are, politicians are more likely than you are to consider opposing points of view because they have to defend their positions in public. And, since their livelihoods depend on the people who vote for them, politicians are also more likely to change their opinions based on the views of their constituents.

If you're regularly **consciously biased** you're a loner, even when you're in a group. The groups are often "hate groups", or at least "seriously dislike groups", in which the participants view people with different perspectives as their enemies. Very similar people are in "love groups" in which something about them, such as their lifestyle, is the only way everyone should be. Since these people are consciously biased about everything, they disagree on most topics and restrict their group activities to their one common group. Keep in mind that just because you've got a conscious bias about one topic (such as your religion or your political party) it doesn't mean you fit in this category. Also, people in certain occupations, such as salespeople, are always biased towards their product or service as opposed to competing products or services, but they don't necessarily bring that bias into other parts of their lives. People with a consciously biased level of fairness are consciously biased about almost everything.

If you're **superficially biased** you're a grumpy individual that others tend to avoid. You're not a bad person, just one who never seems to like anything. If someone buys a birthday present for you there will be something wrong with whatever it is.

You might wonder why people with a **superficially impartial** level of fairness have an evaluation score equal to those who are superficially biased, and lower than those who are both biased and impartial. That's because you can't rely on the opinions or attitudes of superficially impartial people and too often see them lose what little fairness they've shown. Superficially impartial people just don't invest enough effort in being impartial to have it be part of their personality. People who are both biased and impartial are, at least, consistent on each issue.

If you're **consciously impartial** you carefully follow the principle that decisions are based on objective criteria rather than on personal preference. All points of view relating to the topic are considered and then the best is chosen. If this sounds familiar, that's because it's a description of how scientists form conclusions using the scientific method. Don't think for a moment that scientists are always consciously impartial – generally, when they propose a theory, they have a real bias as to the expected result. The key point is that by following the scientific method you force impartiality on your results. After all, someone else with opposing views can perform that same experiment, and if they come up with different results because you were biased, your scientific career is in trouble.

You'll also be a good leader because others see you as someone they can trust to make fair decisions, even if those other people don't agree with the decision. Being consciously impartial isn't easy. As the term implies, you have to consciously consider various points of view and use impartial judgment to make decisions. Just the act of considering other viewpoints is a form of recognition that transmits a positive acknowledgement to the people who presented the ideas, even when their ideas are rejected.

Improving Your Fairness

Before you can improve your fairness you have to recognize if you have a bias when considering other points of view. Only family members and very close friends will know you well enough to see a general bias in your personality. Other people may see a bias once in a while, but nobody's perfect.

If you're **consciously biased** you probably have a long road to travel to become **impartial**. Start by finding people who are well respected in your community or social group but with points of view with which you disagree. Bring yourself to realize that such respected people must not be idiots. Once you've gotten that far, opening your mind to their ideas will be easier. This is not to say you'll accept their ideas, just that you recognize them to be valid possibilities. Continue this process with as many people as possible. Of course, you'll also be considering opinions from other consciously biased people, and recognizing their bias may help you recognize your own.

If you're **superficially biased** you probably know you're considered grumpy. Changing your attitude will help the people you love love you more and will help you realize that life is too short to be grumpy all the time. Crack a smile, even if it hurts a little at the start. Try to relax and let others do the criticizing for awhile. You should find people will relate to you better the more you share with them and the less you complain to them, and you'll have more opportunities to participate in things you like.

If you're **superficially impartial** you're seen as someone who isn't fair all the time even on the same issues. Since others distrust your attitude it's hard for you to maintain close friendships. The solution is clear – consider what your opinion is on an issue and stick with it. If you do it by becoming more biased you'll have those problems to correct later. If you do it by becoming more impartial you'll see more pleased faces on others when you're involved in an issue because they know your decision will be fair, and you'll be on the way to a **consciously impartial** level of fairness.

If you're both biased and impartial, it's better if others know on which issues you'll be biased and on which you'll be impartial. If that's the case, use the evaluation number shown above. If they don't know when you're going to be biased or when you'll be impartial, give yourself a 1 and try to figure out why you waffle so much. It might be something as simple as not taking the time to learn the issues, and when you're unsure of what you believe, as Albarracin found, you take the biased approach of reverting to your previous position. Developing more confidence in your decisions should help to make them more impartial.

Neuroticism
Tendency to Experience Negative Emotions

Emotional Stability

The extent of your patience when frustrated by difficulties. Range: **determined** to **panicky**.

Your Essence Traits

Evaluation:

Determined	= 7
Calm	= 10
Average	= 5
Sensitive	= 4
Angry	= 2
Panicky	= 0

Emotions are the most obvious aspects of a personality, a kind of window to the inner person. People who are always full of emotions are seen as high-maintenance individuals with whom relationships would be difficult. Their emotional stability is nonexistent. On the other side, people who exhibit few emotions are seen as bland individuals with whom relationships would be difficult. Their emotional stability is so steady they never seem to be interested in anything. If your computer exhibits more emotions than you do, you should either start taking real interest in the activities around you or get a calmer computer.

There are two general kinds of emotions: those directed at objects and those directed at people. Emotions directed at objects include disgust, fear, and appreciation of natural beauty, and your approach to them is whatever is natural for you. No problem. Those are also emotions directed at people, but more important ones include anger, gratitude, shame, and romantic love. All of them are much more complicated because your emotions are generally directed towards their emotions and, as Steven Pinker notes, "The problem in dealing with people is that people can deal back."

As you can see in the evaluation numbers, you're the most emotionally stable when you're **calm**. In other words, *don't panic*. This commonly given advice has always been the right approach when under stress. If a soldier in a battle situation panics, he's much more likely to be killed. If a job applicant panics during the interview, she'll never get the job. If a student panics when taking a test, his brain often goes blank. Don't panic.

Calmness is underrated. The period of Theta waves, which occur when you're drowsy, is often the time when your best ideas come to you.

What happens when you get emotional? Sometimes it seems that your emotions take over and present a version of you that isn't you. Emotions can trigger memories and behaviors that have you act differently than you expect to or want to. But to a large extent, you choose the emotions you have. Going for an important job interview? A confident emotion is better than a despondent one. Helping a friend bring their sick child to the hospital? If your emotions are happy and carefree, you may have just lost a friend.

You should do an occasional emotional check on yourself. Is the emotion you're feeling a true reflection of your current situation? To answer that, you have to understand your current situation. Did you just watch a movie about your favorite superhero saving the world and now you feel all powerful? This is not the time to tell that rough looking group across the street to get lost. Did you see a news report about starving children in another part of the world? Contacting an aid agency or your political representative to ask how you can help is a good response (certainly follows the Golden Rule), but becoming depressed about the plight of those children isn't. Both of those scenarios don't describe your current situation – they're situations for other lives that you've temporarily taken for your own.

If you're having a bad day and are feeling negative emotions (anger, despair, fear, frustration, insecurity, disappointment, discouragement, etc.), just telling yourself you're having a bad day may help you choose an emotion that makes the rest of your day run more smoothly. But you might be feeling those emotions for very good reasons, and then those emotions should be kept and exhibited. Many times a negative emotion will be the starting stimulus for the individual to change. People disgusted with their weight start a diet; people disgusted with their menial jobs go back to school; athletes disgusted with their performance practice that much more; people disgusted with their society are the ones who start social changes. Negative emotions can be very good stimuli for change, but they rate rather low when evaluating emotional stability.

Very often, though, a negative emotion is directed at another individual, not at a situation in your life. You may believe this other person is keeping you from having something you want (food, sleep, a relationship, etc.). If that's true, your emotion is correct. However, you could be mistaken or the other person may not realize they are inhibiting you, in which case you'll be viewed as someone who jumps to conclusions – a kind of emotional instability. Look before you leap.

Positive emotions (love, confidence, friendship, curiosity, humor, peace, kindness, etc.) are great, but if you choose such an emotion to gloss over your true feelings the people around you will get an incorrect view of how you really feel. Your true feelings will exhibit themselves in one way or another, and by masking your emotions you'll be seen as emotionally unreliable. And recall that I mentioned in Chapter 7 that friendships are lost and even wars are started because emotions for small problems have been suppressed, and that the best way to keep things stable is the *tit-for-tat* method.

What signs do you look for in evaluating emotional stability? Naturally, if you have a history of nervous breakdowns, your emotional stability needs help. Do you drink excessively, change your job frequently, always have to be right, or can't keep friends? You may have emotional stability issues and would benefit from consulting a professional.

Improving Your Emotional Stability

We're told that being emotionally unstable is undesirable, that we should avoid people who appear unstable, and that we should strive to control ourselves in the same way. But you probably know people who are so controlled that others make jokes about it. Here are three guides to achieving a sensible level of emotional stability:

Your Essence, Your Eternity

- **Don't worry about things you can't change.** If you're outside playing tennis and it starts to rain, screaming at the clouds will make your tennis friends run for cover (from you, not the rain). If you get a flat tire and don't have a spare, a little improper language may be OK but deciding on the most sensible thing to do demonstrates proper emotional stability. If your boss has always been a louse, he'll stay a louse, and all you can do to maintain your own emotional stability is to accept that or quit. If your daughter is on her first date, think about the wonderful time she's having and how mature she's becoming rather than fretting about you-know-what.
- **Focus on the present, with an eye to the future.** Why is it that people don't seem happy with what they have and are always trying to get more? This seems to start in the teen years, when a 13-year-old wants to be 16 and a 16-year-old wants to be 18 and an 18-year-old wants to be 21. Often those teenagers try to give the impression they are older, but what happens is not what they expected – they don't get to enjoy the things directed to their younger age and

eventually find out that being older isn't as great as they thought it would be. The ultimate result is a feeling of loss which, being teenagers, they blame on their parents. Of course you should look to improve yourself and your situations for the future, but don't forget to enjoy what you have now. If you only look to the future you'll be unhappy with the present.

The Canadian comedian Red Green talked about age like this:

> Teenagers sometimes get fake IDs to show they're older. When I turned 56 I did the same thing – I got an ID showing that I was 74 years old. Now when someone asks for my ID they always say, "Wow! I hope I look as good as you do when I'm as old as you are." He feels good for giving me a compliment, I feel good for looking so young, and we both leave happy.

- **Know thyself.** This ancient Greek aphorism still applies today. Try to find out how other people see you. Most of those views should be independent and fair. If you don't like what you're hearing, consider making changes in yourself to improve. The process improvement method described in Chapter 14 will work here, too. As this process implies, don't see yourself as someone you're not. There are many problems in the world and you can only fix some of them. If you're a teacher, you shouldn't spend sleepless nights because a cure for cancer hasn't been found. You can support causes you believe in, which will help your happiness level, and you can encourage people in authority to help, but if you don't recognize your own limits you'll always be unhappy.

You may think that emotionally stable people are boring. It's more accurate to say that boring people are usually emotionally stable. What's most important for your emotional stability is to be able to calmly evaluate difficult situations and take the best course of action in the circumstances you have. Don't panic.

Apprehension

The level of confidence you have. Range – **extremely insecure** to **over-confident**.

Your Essence Traits

Evaluation:

    Extremely insecure = 0
    Insecure = 3
    Neutral = 7
    Self-assured = 10
    Over-confident = 0

Everyone's self-assured about situations they know well – either they're sure they can do it or sure they can't. The apprehension personality attribute isn't about known situations; rather it concerns your attitude towards new situations.

Do you feel insecure when participating in something new? Insecure people often become perfectionists and high performers as the best way to limit their chances of failure. What turns out to be a terrific work ethic really results from a negative personality attribute. If you're extremely insecure it can paralyze you to the extent that you may avoid the situation entirely and, if it involves danger, you might find yourself in a "deer in the headlights" position with possible dire consequences. If you see this happening to you frequently, you could become unhappy with your life.

Being over-confident is no better. How many times have you seen experts in some activity (sports, science, marketing, etc.) do or say something that's so bad you wonder how they could have done it? Over-confidence is a symptom of larger problems that include complacency, arrogance, and outright denial of their faults. When things do go wrong, the over-confident person blames others. Not a good way to form relationships.

You see that extremes in apprehension, whether due to insecurity or to over-confidence, are self-destructive. The best attitude is to be self-assured with yourself and what you can and cannot do. This requires some introspection and a relatively calm approach to new situations. Certainly, some situations will make you feel insecure, and there's nothing wrong with that. Being self-assured with yourself in general will limit your insecurities and permit you to focus on important specific aspects of the situation rather than becoming overwhelmed by the whole thing.

Your Essence Traits

Improving Your Apprehension

Before you can improve your apprehension attribute you have to recognize your strengths and weaknesses. To be as honest as possible, pretend you're someone else and you're describing this other person you happen to know very well (you). Make two lists, one for the strengths and one for the weaknesses of that person.

Now look at each item on these lists. Since you're trying to become more confident without becoming overconfident, look with pride at your strengths and then note weaknesses in those strengths. Even someone as perfect as you can find some weaknesses in your strengths. For example, there's a story about a poll that was taken in which top athletes were asked to name the best athlete in their sport. Each of them chose themselves. The poll was repeated in which the athletes were asked, "After you, who is the best athlete?"

Now look at your list of weaknesses, without becoming depressed, and find strengths in those weaknesses. The goal here is to make your strengths even stronger by eliminating weaknesses within them but recognizing that weaknesses will always be there. In the same way, improve your confidence by eliminating the general weaknesses you have, recognizing that there will always be some strengths within them.

Tension

The level of your patience. Range – **comatose** to **demanding**.

Evaluation:

Comatose	=	0
Relaxed	=	10
Neutral	=	8
Anxious	=	5
Demanding	=	0

What exactly is patience? A dictionary definition is "the capacity to endure waiting, delay, or provocation without becoming angry or upset." People who are patient are tolerant of life's little annoyances, lenient towards the people who annoy, and reasonable towards the requests of other people. A patient person isn't the same as a meek person – a patient person is comfortable with their life while a meek person is afraid of their life.

Your Essence Traits

Consider dogs. A Bernese mountain dog, which is a very large animal, will quietly stand still while a tiny Papillion is barking his head off at it. Throughout the animal kingdom large animals exhibit an almost unlimited amount of patience when bothered by smaller animals (of course, threatening interactions such as jackals surrounding a cape buffalo are treated differently). When you're big, it's easier to be patient.

Everyone wants you to be patient except the people to whom you owe money. Patience is extolled in all the major religions as one of the most important practices to have, and it's listed as one of the *Seven Heavenly Virtues* promoted in some religions. How many times have you heard, "Rome wasn't built in a day", "all in good time", all things come to him who waits", "don't hold your breath", and "time will tell"? When comparing the relative values of intelligence and patience, a Dutch proverb says, "A handful of patience is worth more than a bushel of brains." So why are people so impatient?

Sometimes your patience is tried. If your wife (or husband – no sexual clichés here) announces that you're going to the opera tonight and you hate opera, you'd better know how to be patient to reduce your tension. If your husband (or wife) announces that you're going to the football game and you hate football, your patience abilities will be called upon. If you're sitting in a traffic jam and start blowing your horn, you might learn patience from a punch in the nose by another driver (that other driver should improve their patience, too). Patience is, after all, a virtue. Being virtuous takes patience.

But you don't want to be *too* patient. Being **comatose** certainly means you have no tension associated with your personality, but it also implies that you don't pay attention to tense situations. In this case, of course, "comatose" doesn't mean actually being in a coma. It means being so calm about everything that people may think you're in a new kind of walking coma.

Anxiety is an important trait for animals in the wild and was an original trait for humans. Being anxious is what kept us vigilant and alive in a world where being someone's next meal was the norm. Anxiety also keeps us alert to climate excesses, aggressive strangers, diseases, potentially harmful situations ("Don't run with a scissors"), and things that go bump in the night. The key is to maintain a healthy level of anxiety while not being so worried about things that might happen that you become obsessively fearful. Being scared is a limitation on your life and being afraid of everything (*panophobia*) is hardly a life at all.

Being **demanding** suggests that you're internally tense, but what it certainly means is that people interacting with you become more tense. Those other people find their patience stressed, possibly resulting in a confrontation. Demanding people are avoided and often shunned. One of the quickest ways to lose the few friends you have is to start demanding things from them.

The most desirable level of tension for yourself and for everyone around you is being **relaxed** during tense situations. Relaxed doesn't mean uncaring – it means you assess the best approach to the situation without yelling and without panicking. You exhibit the appropriate level of patience with the situation and with everyone associated with it. There are certainly some situations that require confrontations, such as discovering that a trusted individual has been stealing from you. But for the many tensions we encounter in our day-to-day life, developing a relaxed attitude is the best personality trait for you.

Improving Your Tension

Here are some suggestions for improving your patience:

- Recognize what triggers your impatience, especially just before you blow.
- Experiment with ways to reduce your impatient feelings, such as taking quiet breaths or telling yourself "it doesn't matter".
- Find ways to slow down. Impatience comes from wanting to do things quickly.
- If you're becoming impatient over a ridiculous situation, laugh at it.
- Think about all that you've accomplished rather than all that you still need to do.
- Count to 10. If that doesn't work, count to 100.
- Daydream about being in a very peaceful, happy place. Don't do this while driving.
- Excessive amounts of some stimulants, such as caffeine, can make you edgy.

Egocentrism

The extent of your focus on yourself. Range – **dangerously egocentric** to **servile**.

Your Essence Traits

Evaluation:

Dangerously egocentric	= -10
Arrogant	= 0
Vain	= 3
Assured	= 10
Humble	= 7
Servile	= 1

How self-centered are you? Jean Piaget thought this question should only be asked about people older than 15 years because young children are normally egocentric. Most parents would agree and most parents of teenagers would wholeheartedly agree. Consequently, this discussion will focus on those people who have fully developed personalities. Or who should have, anyway.

If you're considered **dangerously egocentric** you have what psychiatrists have classified as a *Cluster B personality disorder*, and it could be *antisocial, borderline, histrionic,* or *narcissistic*. Not something to be proud of. But if you're dangerously egocentric you're proud of everything about you, so let's describe your personality disorder more completely. These descriptions are from the *DSM-IV* (*Diagnostic and Statistical Manual of Mental Disorders* 4th ed., 1994), published by the American

Psychiatric Association. If you're one of these, do something about it.

- **Antisocial Personality Disorder.** These people have been called sociopaths or psychopaths and are the serial killers in our societies. In this case you have a pervasive pattern of disregard for and violation of the rights of others occurring since age 15 years, as indicated by three (or more) of the following:
    - Failure to conform to social norms with respect to lawful behaviors as indicated by repeatedly performing acts that are grounds for arrest;
    - Deceitfulness, as indicated by repeated lying, use of aliases, or conning others for personal profit or pleasure;
    - Impulsivity or failure to plan ahead;
    - Irritability and aggressiveness, as indicated by repeated physical fights or assaults;
    - Reckless disregard for safety of self or others;
    - Consistent irresponsibility, as indicated by repeated failure to sustain consistent work behavior or honor financial obligations;
    - Lack of remorse, as indicated by being indifferent to or rationalizing having hurt, mistreated, or stolen from another.

- **Borderline Personality Disorder.** Borderline doesn't mean borderline good; it means borderline

Your Essence Traits

bad. While antisocial people use their power to manipulate others, borderline people use their weaknesses to do the same thing. In this case you're close to psychotic and you have a pervasive pattern of instability of interpersonal relationships, self-image, and marked impulsivity beginning by early adulthood and present in a variety of contexts, as indicated by five (or more) of the following:

- Frantic efforts to avoid real or imagined abandonment;
- A pattern of unstable and intense interpersonal relationships characterized by alternating between extremes of idealization and devaluation;
- Identity disturbance: markedly and persistently unstable self-image or sense of self;
- Impulsivity in at least two areas that are potentially self-damaging (e.g., spending, sex, substance abuse, reckless driving, or binge eating);
- Recurrent suicidal behavior, gestures, threats, or self-mutilating behavior;
- Affective instability due to a marked reactivity of mood (e.g., intense episodic dysphoria, irritability, or anxiety usually lasting a few hours and only rarely more than a few days);
- Chronic feelings of emptiness;
- Inappropriate, intense anger or difficulty controlling anger (e.g., frequent displays of

temper, constant anger, or recurrent physical fights);
- Transient, stress-related paranoid ideation or severe dissociative symptoms.

- **Histrionic Personality Disorder.** Histrionics are the drama queens/kings of the world. In this case you have a pervasive pattern of excessive emotionality and attention seeking, beginning by early adulthood and present in a variety of contexts, as indicated by five (or more) of the following:
  - Is uncomfortable in situations in which he or she is not the center of attention;
  - Interaction with others is often characterized by inappropriate sexually seductive or provocative behavior;
  - Displays rapidly shifting and shallow expression of emotions;
  - Consistently uses physical appearance to draw attention to self;
  - Has a style of speech that is excessively impressionistic and lacking in detail;
  - Shows self-dramatization, theatricality, and exaggerated expression of emotion;
  - Is suggestible, i.e., easily influenced by others or circumstances;
  - Considers relationships to be more intimate than they actually are.

Your Essence Traits

- **Narcissistic Personality Disorder.** In this case you have a pervasive pattern of grandiosity (in fantasy or behavior), need for admiration, and lack of empathy, beginning by early adulthood and present in a variety of contexts, as indicated by five (or more) of the following:
    - Has a grandiose sense of self-importance (e.g., exaggerates achievements and talents, expects to be recognized as superior without commensurate achievements);
    - Is preoccupied with fantasies of unlimited success, power, brilliance, beauty, or ideal love;
    - Believes that s/he is "special" and unique and can only be understood by, or should associate with, other special or high-status people (or institutions);
    - Requires excessive admiration;
    - Has a sense of entitlement, i.e., unreasonable expectations of especially favorable treatment or automatic compliance with his or her expectations;
    - Is interpersonally exploitive, i.e., takes advantage of others to achieve his or her own ends;
    - Lacks empathy: is unwilling to recognize or identify with the feelings and needs of others;
    - Is often envious of others or believes that others are envious of him or her;
    - Shows arrogant, haughty behaviors or attitudes.

Your Essence, Your Eternity

Without being dangerously egocentric, individuals express themselves at some egocentric level. Keep in mind that egocentric isn't related to *ego* as defined by Sigmund Freud. In this case, your egocentric level is basically how you feel about yourself.

If you have an **arrogant** egocentric level you let others know that you're better than them in some way. This is certainly not the best way to get along with others, but some people act this way anyway to feel superior and to get the red carpet treatment. Well, if you're going to be arrogant you might as well be good at it, so here are some tips:

- Believe you're as good as you say you are. Dismiss any nagging doubts.
- Have a silent presence and aura.
- Smile infrequently.
- Walk with long strides.
- Don't have any gaps in your area of superiority. Gaps give others a chance to question you. If they do, imply they don't understand.
- Don't be rude. Use good manners arrogantly.
- Dress well.

If you have a **vain** egocentric level you focus on me, me and me. You want to be sure others know how wonderful and perfect you are, so you should stop holding back on talking about yourself. As Isaac Asimov said

Your Essence Traits

(tongue in cheek), "People who think they know everything are very annoying to those of us who do."

Also be sure everyone knows how you intimidate both men and women, physically and mentally. If there's something about your physique that's too small or too big, the exercise people, pill people, and cosmetic surgeons are ready to help. The difference between an arrogant person and a vain person is the sneer (arrogant) versus the superior smile (vain). Your smile will show everyone you're not arrogant.

If you have a **servile** egocentric level you're marginally better than someone who's arrogant, but being servile is not a good way to live. You may think you're just listening to the advice and opinions of others, but servility is more like being a slave than being flexible. You bow to the direction of most people you meet, often to avoid the conflict you're certain will come if you don't. Others recognize this weakness and take even more advantage of you, testing your servility limits. If you're a naturally servile person, here's how to make the best of it:

- Pick a good dominant person to serve. That will, at least, extract you from the "servile pool" that all the pushy people use to get temporary slaves.

- Set some limits with your dominant. Neither your dominant nor you want you to start lashing out because you've been pushed too far.
- A dominant/servile relationship isn't about sex – it's about control.
- You don't have to be perfectly servile all the time because that's boring to your dominant. Being reluctant from time to time permits your dominant to assert their power. However, you do have to obey almost all the time.
- Say goodbye to jealousy. A servile person can't be jealous of anything their dominant does. If you do get jealous, share it with your dominant so there are no hidden frustrations.
- Respect your dominant. They expect it.

If you have a **humble** egocentric level you have a gentleness of spirit that others see. Being humble isn't easy, especially in a competitive society. By the way – you can't be competitively humble. As country musician Mac Davis said:

> Oh Lord it's hard to be humble when you're perfect in every way.
> I can't wait to look in the mirror 'cause I get better lookin' each day.
> To know me is to love me, I must be a hell of a man.
> Oh Lord it's hard to be humble, but I'm doin' the best that I can.

Your Essence Traits

Besides being something almost every religion encourages, being humble has immediate benefits for the individual. You make a conscious effort to ask people about themselves rather than only talking about yourself, which encourages others to be more open with you. If you don't brag about your accomplishments you won't have to explain your failures. If you're humble you know you don't know it all and you're more interested in personal development and learning in general. You're also more honest with yourself.

Having an **assured** egocentric level means you've taken the positive characteristics of being humble and enhanced them with a level of self-confidence that drives your success. By removing all appearance of arrogance, vanity, and servility from your interactions with others, you're someone who's respected and often chosen to lead activities. The hardest thing about being assured is to avoid dropping into a vain egocentric level by becoming so proud of getting all that attention from others.

Improving Your Egocentrism

If you have a **dangerously egocentric** personality, run, don't walk, to a licensed psychiatrist. There's no assurance you can improve yourself without professional help.

If you have an **arrogant** egocentric level you can improve yourself by reversing some of the bullet points describing that characteristic. For example, you can

- Not believe you're as good as you say you are
- Contribute more in conversations and activities
- Smile frequently

If you have a **vain** egocentric level the easiest way to improve yourself is to stop talking about yourself. Completely. That includes any activities or interests you have and anything else that refers to you. Instead, talk about someone else's activities or interests (no gossip, please) or, if that's too hard, talk about the weather. Talk about anything except yourself. If another person says something complimentary about you, say a simple "thank you" rather than embellishing their compliment with more compliments about you from you. In one of those amusing paradoxes of life, if you succeed in curing yourself of being vain you'll have something to be really proud of. However, you've cured yourself of this kind of vanity, so you won't be proud.

If you have a **servile** egocentric level, the best approach is the *tit-for-tat* method, described in Chapter 7 under *The*

Your Essence Traits

Golden Rule In Practice. While tit-for-tat is effective, it requires a level of confidence that servile people may not have. Have a friend practice tit-for-tat scenarios with you to help build your nerve.

If you have a **humble** egocentric level all you need to do to improve is to become more self-confident in your abilities. While you're doing this, concentrate on maintaining that wonderful humble nature that's made you the nice person you are. Self-confidence and humbleness are a bit incompatible and having them both will require your careful attention.

Here are some qualities you can develop to have a more assured level of egocentrism:

- **Appreciate your talents.** Assured people feel good about themselves, with a healthy self-esteem. This is different from pride, which is projected in a more arrogant way and indicates a level of insecurity.
- **Conduct an honest evaluation of yourself.** If you find a weakness in yourself before other people do you'll be able to correct it before they exploit it.
- **Recognize your faults and limitations.** No matter how good you are, someone else is better. You can improve yourself by learning from others who do something better than you.
- **Stop comparing and appreciate who you are.** Being assured is also being humble and comparisons

with others encourage unhealthy competitions to prove who's "better".
- **Don't be afraid to make mistakes.** Everyone makes mistakes. When you make one, admit it.

By now you should understand a lot more about yourself than you did before reading this chapter. But it's not enough – your essence traits interact with your culture and habitat and you have to understand those interactions before you can really understand yourself.

Chapter 10 – Your Essence Interactions

While the personality attributes define the essence, the essence interacts with the other components of living things (see Figure 1-1) in complex ways to define the complete individual. Different personality attributes relate to different components, each affecting the other. Those relationships are shown in Figure 10-1, with attributes grouped as the *Big Five Personality Factors* of Openness, Conscientiousness, Extroversion, Agreeableness, and Neuroticism. *Openness* measures imaginative vs. practical behavior and includes a person's appreciation for art, emotion, adventure, unusual ideas, curiosity, and variety of experience; *Conscientiousness* measures planned vs. spontaneous behavior and includes a person's tendency to show self-discipline, act dutifully, and aim for achievement; *Extroversion* measures the level of engagement with the external world and includes a person's energy, positive emotions, and the tendency to seek stimulation in the company of others; *Agreeableness* measures compassion vs.

antagonism and includes a person's tendency to be compassionate and cooperative rather than suspicious and antagonistic towards others; and *Neuroticism* measures the tendency to experience negative emotions and includes a person's anger, anxiety, depression, and vulnerability.

You'll notice that some attributes appear under more than one component, illustrating the broader nature of their impacts. Actually, most of the attributes will affect all of components of living things in some way, but those listed under a component in Figure 10-1 have the greatest effects.

The single guidance for improving your essence is the Golden Rule, "Do to others as you would have others do to you."

Big Five Domain	Physical Body	Culture	Habitat
Openness	Philosophical Attitude	Liveliness Abstractedness	Liveliness Abstractedness
Conscientiousness	Maturity	Task Performance Attitude	
	Achievement Attitude	Control Attitude Vigilance	
Extroversion	Warmth	Material Attitude	Material Attitude
	Socialization	Risk Attitude	Socialization
	Risk Attitude	Privateness	Risk Attitude
	Leadership		
Agreeableness	Sensitivity	Sensitivity	
	Aggressiveness	Agressiveness	
		Ethics	
		Morals	
		Fairness	
Neuroticism	Tension	Tension	Emotional Stability
	Apprehension		
	Emotional Stability		

Figure 10-1. **Personality Attribute Interactions**

Essence Interaction with the Physical Body

The connection between the physical body and the essence has been an active topic among philosophers for millennia and has more recently regained interest in the scientific community. The range of thought, as you might expect, is wide: some researchers believe everything is biochemical and there is no ethereal aspect, while others go the opposite way with the physical body being an illusion of the essence energy. William Tiller expresses some of the main precepts of the essence when he says:

> My working hypothesis is that we are all spirits having a physical experience. To effectively have this experience, we need a suitable structural interface with the spacetime world. That became a *biobodysuit* constructed from the substance complex presented by equation 5 {a mathematical description of the physical body as a complex of Gaussian-shaped particles and waves in space}. That type of biobodysuit is what we put on when we are "born" into spacetime and it is what we take off when we appear to die in spacetime. In between, when we are manifesting what we call "life," this biobodysuit contains what I label our *personality* self. (p. 309)

Personality attributes, by default, operate from the brain, a part of the physical body. Therefore, with both the brain and the essence contained in the same physical body component, this interaction is the most direct among all the components of living things. Some researchers have noticed relationships between personality and other parts of the physical body. For example, Mats Larsson and his colleagues have found that people with more pits in the iris of their eyes are tender, warm, and trusting, while people with more lines curving around the outer edge of their iris are more likely to be neurotic, impulsive and give in to cravings.

Here's what each personality attribute that affects the physical body contributes to this interaction.

Openness

Imaginative vs. Practical people

Philosophical Attitude is how strongly you adhere to your point of view, from being completely set in your ways to frequently changing your opinion on the same topic.

People have always been ready to help you believe that what they have/want/do/think is the right way to have/want/do/think. Your willingness to listen to different perspectives and make intelligent judgments based on moral and ethical considerations (other attributes of the essence) determine how you interact with other parts of life and with the world. If your judgments are to continue with your original line of reasoning because you've minimized others' concerns, your judgments are not intelligent. Blind stubbornness not only freezes your philosophical attitude, it also inhibits your life and fossilizes part of your essence, making evolutionary growth that much more difficult. It also inhibits your intellectual growth within your physical body because blind stubbornness is totally contrary to the scientific method. Scientists are always ready to form new theories based on new evidence, a mark of a flexible philosophical attitude.

Conscientiousness
Planned vs. Spontaneous Behavior

Maturity is how much you consider consequences in making decisions, from being infantile to making choices wisely.

Maturity is mostly directed from the physical body to the essence than the other way around. Certainly, poor judgments can negatively affect the physical body, such as jumping off a roof in an attempt to fly. And good judgments can positively affect the physical body, such as not smoking. But most of the consequences for this attribute come from the physical body because the physical maturity of the brain is an ongoing process from before birth to past adolescence.

One aspect of maturity reflects a person's approach to knowledge. For example, suppose two ladies are listening to classical music. One of them knows the name of the piece, the composer, and characteristics that make this music particularly important in her culture. The other lady doesn't know anything about the music she's listening to. The second lady would be immature if she pretended to have such knowledge or dismissed the knowledge as not worth having. And the first lady would be immature if she acted superior to her friend because of the knowledge she possessed. In another example, if a man is eligible to vote but never bothers to learn the issues that define the candidates, he is immature. He has not considered the consequences of his voting from ignorance.

Achievement Attitude is the extent of your motivation to reach goals, from reluctant to obsessive.

In a study from 1967 to 1977, a team of epidemiologists led by Sir Michael Marmot of University College London monitored the health of more than 17,000 civil servants in Britain. They found that the higher a person had been promoted, the better their health was and the longer they lived. In particular, mid-level civil servants were more likely to develop heart disease, high blood pressure, lung cancer, and gastrointestinal problems than persons promoted to work directly for cabinet ministers. Typical poor health behaviors such as smoking, excessive drinking, or lack of exercise were ruled out. As Marmot said, "The higher the social position, the longer people can expect to live, and the less disease they can expect to suffer." So, if you want a nice long life, develop an achievement attitude that will reach your goals.

Extraversion

Level of Engagement with the External World

Warmth is the extent of your attention to others, from being very distant to being very outgoing.

There have been several studies showing a close relationship between the psychological warmth you feel towards a person and the physical warmth of your surroundings. Harry Harlow gave macaque monkeys a choice of a wire surrogate mother that provided food or a cloth surrogate mother that provided warmth (a light bulb was behind the cloth) and the monkeys preferred the cloth mother. Also, monkeys "raised" by the warm cloth mother showed relatively normal social development as adults, in stark contrast to the infants left alone with the wire mother.

In a study by Williams & Baugh, people were briefly given a cup of hot coffee or iced coffee to hold. To help insure that the participants wouldn't suspect some link between the coffee and the project, a confederate who didn't know what the study was trying to do met participants in the lobby carrying a cup of coffee, a clipboard, and two textbooks. During the elevator ride to the laboratory, the confederate casually asked participants if they would hold the coffee cup for a second while she recorded their name and the time of their participation. After the confederate wrote down the information, she took back the coffee cup. The temperature of the coffee cup (hot versus iced) was the only between-subjects manipulation. Those people who held the hot coffee judged a target person as having a "warmer" personality while those who held iced coffee judged the target person to have a "colder" personality. In another study, people who briefly held a hot therapeutic pad (supposedly for product evaluation) were more likely to buy a gift for a friend; those who held a cold pad were more likely to buy a gift for themselves.

What would cause such a physical/psychological relationship? Further investigations have found that the insular cortex, a small section of the brain previously ignored but now found to be involved in everything from telling time to sexual interest, is also the place where feelings of warmth (both physical and psychological) are interpreted. So, if you want to have a warmer relationship with someone, give that person something warm to hold.

Risk Attitude is the extent of your willingness to take chances, from being guarded to being reckless.

It's not "bad" to be so safety conscious that you hold onto two railings when climbing stairs, nor is it bad to be so unconcerned that you skydive to relax from your more extreme activities. By the way – neither one is "good", either. From a physical body perspective, your need for risk is based on your biochemistry (such as testosterone level) modified (or not) by fear or prudence factors. An example of a person with a high-risk attitude was the daredevil motorcyclist Evel Knievel who broke so many bones during his jumps that, when he was forced to retire in 1980, told reporters that he was "nothing but scar tissue and surgical steel."

Some activities are designed to provide thrills without the constant fear of death or serious injury. For example, roller coasters give a sense of danger without actually causing injury (usually), and the extent of that sense of danger is designed into the ride. Some people, with low levels of risk attitude, refuse to ride any of them. You can think of many examples of dangerous activities, such as bungee jumping, mountain climbing, and walking through some parts of many big cities.

Finally, we consider those people who handle poisonous snakes (or other dangerous wildlife) without protection, those who play Russian roulette, and other similar activities. In these cases, foolishness outweighs maturity or even common sense. Sometimes called the *Darwin Effect* because of its likelihood to remove such individuals from the gene pool, ultra-extreme high-risk attitudes should be identified and eliminated from your personality, not because of negative effects to your essence, but because of their traumatic (at least) effect on others who happen to be nearby.

Leadership is the extent of your interaction in a group, from being completely independent to being both a follower and a leader.

You're considered an effective leader if you exhibit other attributes associated with the physical body, specifically emotional stability, maturity, risk attitude (sensible, of course), aggressiveness, and socialization.

Other leadership qualities include a specific skill, charisma, dedication, results-oriented, cooperation, optimism, ability to delegate, avoidance of bureaucracy, empathy, integrity, and a good sense of humor.

While larger people are often thought of as leaders, leadership qualities seem to override size in choosing leaders. We all know smaller people who are popular with everyone and who just exude leadership qualities. But height comparisons are still done – from 1912 to 2016 the taller candidate has become President of the United States in 20 of the 27 elections (two of them were ties). And, from the old cliché, the taller boy gets the girl.

Of course both males and females can be leaders, but from a physical body perspective they exhibit different behaviors in their leadership. Brain scans show that men focus tasks on one side of their brains, exhibiting qualities of "strong vision" or "results orientation". Women have larger dendrites connecting their brain halves and use both sides of their brains in task accomplishments. Consequently, women have superior multi-tasking abilities and are more detail oriented. The testosterone and adrenaline produced by men help with the assertiveness necessary to compete, while the estrogen and oxytocin produced by women help with the nurturing necessary for a group's success. So, if you want to get bottom-line results, get a male leader; if you want to get collaboration, get a female leader. One thing I'm sure to get is bashed as a sexist for reporting these brain scan conclusions.

In summary, certain personality and physical body attributes affect an individual's ability to be a leader. On the other side, individuals thrown into leadership roles can succeed by improving the necessary personality attributes, even if they lack some of the natural qualities associated with leaders. We can even speculate that "born leaders" are really individuals who have learned the essence attributes needed for leadership.

Socialization is the extent of how you fit in with a group, from being anti-social to being actively social.

Kathryn Wynne-Edwards points out that while most of socialization refers to cultural interactions, there are physical body changes that are associated with it. There have been plenty of studies showing that social isolation is bad for your health. Married people are generally happier and healthier. When your spouse dies you're more likely to die, something called the "widowhood effect." Strong social networks help prevent Alzheimer's disease and weak social networks are as great a risk for heart attack patients as are obesity and hypertension. Men who become fathers for the first time show a decrease in testosterone and cortisol levels and a higher level of estradiol concentrations, a hormone known to influence maternal behavior. Your level of socialization affects your physical body more than you think.

External adjustments to the physical body are sometimes done to correct antisocial behavior. A great variety of antidepressant drugs have been developed to treat every level of social maladjustment, but drug treatment only covers the problem without curing it. At the same time, interactions with other drugs or with the internal chemistry of the individual can sometimes cause significant side effects, up to and including permanent psychosis and death. Pharmacology is tricky enough, but when we move towards psychopharmacology we're dealing more with an art than a science.

Psychopharmacology has also been practiced with poorly socialized animals, mostly pet dogs and cats. Wolfgang Jochle notes the three main causes for disturbed or abnormal behavior in animals are the drive for dominance, anxiety and frustration, and geriatric situations.

Hypothyroidism, which is often undiagnosed, is also a common cause for abnormal behavior. Contributing factors are, as any pet owner knows, the breed and sex of the animal. The usual treatment to improve animal socialization is some drug to elevate prolactin levels. Left untreated, the animal may suffer disturbances in the condition of their skin and fur and may result in an immune deficiency.

Of course, antisocial behavior is a main problem from illicit drug use. This is especially curious since a main reason for using such drugs comes from socialization peer pressure. Media reports of people lashing out when they're high on some illegal drug confirm the need to eliminate such drugs, but legal drugs (such as alcohol or antidepressants) can be just as dangerous when used inappropriately. Know thy body.

An extreme example of socialization problems that affect the physical body are men or women who conclude they were born the wrong sex and want to do something about it. The something, of course, is a sex change operation, for which surgical and chemical changes are performed on the physical body to transform the individual physically so they can socialize more comfortably.

Agreeableness

Compassion vs. Antagonism

Sensitivity is the extent to which you're tough-minded or tender-minded, from being completely impersonal to being extremely sympathetic.

Tough-minded people prefer using logic to analyze issues while tender-minded people prefer to use the effects of the various choices on the people involved. You can tell if you're tender-minded by the kinds of entertainment you don't like. Tender-minded people avoid scenes of graphic violence because they continue to remember the scenes as if they were real and they sometimes get nightmares from them, especially if the bad guy is doing the violence. Both tough-minded and tender-minded people can develop nervous or digestive disorders if they think decisions are made contrary to the way they should be (with logic for tough-minded or with social awareness for tender-minded). If you're especially prone to such conditions you should fashion your life to avoid getting involved with them; if that's not possible, some kind of medication (a so-called *happy pill*) may be needed. You shouldn't opt-out of life because some people are doing things with which you disagree, but it doesn't do you or your cause any good if dealing with the issues makes you sick.

Your tough or tender sensitivity level is sometimes reflected in your tolerance of physical pain, both your own and to the suffering of others. People exposed to a terrifying event or ordeal in which grave physical harm occurred or was threatened may suffer *Post Traumatic Stress Disorder (PTSD)*. According to the National Institute of Mental Health, people with PTSD have persistent frightening thoughts and memories of their ordeal and feel emotionally numb, especially with people they were once close to. They may experience sleep problems, feel detached, or be easily startled. The entertainment media tend to ignore PTSD, instead showing heroic actions by individuals who suffer no ill effects from their trauma. There are certainly many examples of heroism and sometimes those people are given the recognition they deserve. However, sometimes even those heroes suffer psychological effects of the events. There's a strong connection between extremes of your essence sensitivity and your physical body.

Like to forget an alarming memory? It turns out the best way is to use *fear extinction therapy* in which you recall the incident that caused the fear. This gives you a window of about three hours to modify it before it settles back into your brain but, unfortunately, it works only with recent memories. But don't be alarmed – relief is on the way. To quiet long-term fears, Johannes Gräff and his colleagues have found that including a drug with fear extinction therapy helps. Well, helps mice, anyway. Human clinical trials are next.

Sensitivity and physical pain are favorite topics of humor. Many people will laugh at pain inflicted on an actor (but not an actress) as long as that pain doesn't seem to really be painful. Examples are the early film comic teams of *Laurel & Hardy, Abbott & Costello,* and *The Three Stooges.* You've probably laughed yourself, and immediately felt badly about it, when you saw someone jump into water they didn't realize was so cold, or trip on ice and fall into a snow bank. No matter what your level of sensitivity, physical pain can sometimes be funny.

Aggressiveness is the manner of your interaction with others, from being predatory to being loving.

This is different from assertiveness, where the rights and dignity of both sides are maintained while you're expressing your point of view. Aggressiveness bypasses such niceties.

Aggressiveness is a complex phenomenon that psychologists have divided into two broad categories: *hostile aggressiveness* and *goal-oriented aggressiveness*. Hostile aggressiveness often has a kind of goal – revenge. Goal-oriented aggressiveness can be hostile, but when it does it's successful only when hostility is needed to achieve the goal. Both begin with an event from outside of the physical body, and both involve the same brain areas. Testosterone typically makes males more aggressive than females, but natural steroid production in the mother after a birth generates maternal aggression. Don't mess with a new mother. As you may know, medical science has produced a whole menu of specialized cocktail drugs to reduce aggressive behavior.

While you might expect that large physical size would correlate directly with increased aggressiveness, that's not the case. Some of the most aggressive warriors in history have been rather physically small men (for example, Alexander the Great and Napoleon). Certainly, a large physical body would provide a level of protection for aggressive behavior and many large people have taken advantage of their size. Similarly, a small physical body provides little protection, which is why very few 5-foot tall, 95-pound high school students physically abuse varsity football players.

The predator/prey situation is an interesting illustration of aggressiveness because the aggressor is not who you think it is. For example, let's say a cheetah is attacking a warthog. The cheetah is reacting to hunger, not aggression. The warthog must make a fight/flight decision, based on its feeling of how successful it would be to stand and fight. If the warthog does decide to fight, it exhibits dramatically increased aggressiveness as a defense mechanism. However, because human behavior is often a complex weave of personal needs mixed with cultural and societal aspects, this relatively straightforward view of predator/prey interactions doesn't directly apply to people.

Neuroticism

Tendency to Experience Negative Emotions

Emotional stability is a measure of your patience when frustrated by difficulties, from being determined to panicking.

Your Essence Interactions

Your emotions are tricky things that can express themselves in non-essence ways. For example, you're invited on a two-hour hike but you're worried that you'll have to use the toilet and there won't be one available. The more you worry about it the more necessary that toilet use becomes. Interestingly, if you didn't think about it at all you would happily enjoy the hike with no problems. Your worry becomes a self-fulfilling prophecy.

A more serious situation occurs when a person believes they're sick when there's nothing the matter with them except an overactive imagination. Psychiatric books are filled with approaches to these *psychosomatic illnesses* which, if the person believes strongly enough that they're sick, will really make them physically sick with their imagined condition. People have gotten high blood pressure, diabetes, arthritis, skin conditions, and many others just by believing they have them. Your emotions can have a powerful effect on your physical body.

Looking at it the other way, emotional stability is affected by body metabolism, especially the effects of hormones and enzymes. Drugs, both prescribed and otherwise, can affect emotional stability temporarily and/or permanently. Your diet also affects your emotional stability, which means that the habitat where your food comes from will affect your physical body and, in turn, your essence. In fact, Leanne Williams and her colleagues point out that just getting older improves emotional stability. On the other side, emotional stability affects physical body attributes such as blood pressure, heart rate, senses, reflex actions, etc. If you become very emotional your blood pressure and heart rate go up. Do you really want high blood pressure? Do you worry about your heart pounding so hard it feels like something will pop? When situations come up (and they will) that cause you great emotional stress, have your body ready for it. Eat a healthy diet, get regular exercise, put some relaxation into your regular life, and you have a much better chance of your physical body dealing with those occasional stresses.

Apprehension is all about your confidence, from being extremely insecure to being over-confident.

This is not as straightforward as it sounds because people sometimes exude confidence when they have none as a kind of protective measure. Everyone is apprehensive at times but some people have found ways to overcome it. There are also different kinds of situations that cause apprehension, such as peer group interactions, social events, performance events, and daily activities. One of the most stressful events in life for some people is public speaking, even something as casual as giving a report to the class in school. The threat of public embarrassment is so great that these people will go to great lengths, sometimes resulting in significant personal sacrifice, to avoid speaking in public. The comedian Jerry Seinfeld expressed it this way:

> According to most studies, people's number one fear is public speaking. Number two is death. Death is number two. Does that sound right? This means to the average person, if you go to a funeral, you're better off in the casket than doing the eulogy.

Apprehension level has a large effect on your ultimate success in many aspects of life. For example, let's say you're a good tennis player and decide to enter a tournament. For your first match your opponent is a world-famous player. Unless you have uncommon control over your apprehension, you've lost the match before you step onto the court. The physical body elements that contribute to your loss include stiffer joints, slower reflexes, less accurate vision, less strength, and less stamina. Quite a variety of physical effects from a psychological cause!

Your Essence, Your Eternity

A different kind of apprehension occurs when you're applying for a job you really want, and when you go for your interview you see a room full of people holding résumés and dressed just like you. Some of them, perhaps many of them, perhaps all of them, appear to be more qualified than you, even though you don't know them and haven't said a word to any of them. If you don't control your apprehension, your chances of succeeding in your interview are slim. These physical body effects include sweating, poorer communication skills, slower analytical skills, and a generally less confident appearance.

Hans Seyle describes a model called the *General Adaptation Syndrome (GAS)* that was developed to describe reactions to stress. GAS consists of three stages and animals in general seem to react the same to them. The first stage is "alarm", which generates a fight or flight response enhanced by adrenaline production. This stage is an immediate physical body reaction to an essence attribute. The second stage is "resistance", in which continued stress causes the body to use resources to cope with it. The third stage is "exhaustion", when the body has run out of convenient resources and now needs extraordinary efforts, such as sweating, increased heart rate, etc. to maintain physical body functions. Long-term effects of stress can include insomnia, ulcers, depression, cardiovascular problems, job burnout, and mental illness. But keep in mind that stress is a very personal condition – what's extremely stressful for one individual can be a dull event to another.

Fortunately, there are established ways to improve your self-confidence and consequently lower your apprehension, and you'll find them in Chapter 10 under *Improving Your Apprehension*. Those skills should be actively pursued – the lower your apprehension the better your essence will be.

Tension concerns the level of your patience, from being comatose to being demanding.

There are physical similarities between apprehension and tension. If you get a headache because you've lost your confidence, it's due to apprehension. If you get an equally painful headache because you've lost your patience, credit your tension level. Physical problems due to tension are more common and more persistent because you have more opportunities to lose your patience than to lose your confidence.

Besides headaches, excessive tension causes muscle pains, cramps, general fatigue, and even panic attacks. Not only are these problems difficult to live with, they can also shorten your life expectancy. If you find yourself with tension pain, do what you can to lower your tension level. You'll find ways to do that in Chapter 10 under *Improving Your Tension.*

Physical problems with tension also exist in non-human animals. Dominant baboons in Africa physically and verbally abuse subordinate baboons as part of maintaining their status. When blood from the subordinate baboons was analyzed, high levels of a hormone associated with stress were found. Dominant rhesus monkeys have lower rates of atherosclerosis than subordinate monkeys, and when dominant monkeys are dropped to a lower social level their rate of heart disease goes up. There's something about being at the top of the hierarchy that gives a physical body more patience with itself.

But tension is both a positive and a negative factor in people's lives and varies greatly depending on the individual. Some people thrive on it and seek out careers and/or personal situations in which success is a demanding adventure. Those people are the leaders of societies, but others with the same characteristics are our leading criminals. To be a leader in anything requires the ability to manage tension.

Essence Interaction with the Culture

Interactions between individuals within a culture not only define the "personality" of the culture but also often modify the personalities of the individuals. Consequently, many of the attributes associated with personality are also associated with culture. As Weber & Hsee have noted:

> ...culture has been found to influence psychological processes and behavior in virtually every study that bothered to examine it as a variable. (p. 612)

Here is what each personality attribute that affects a culture contributes to this interaction. Because nearly all descriptions of culture are for human cultures, interactions with the cultures of other living things are described with whatever information is known.

Openness

Imaginative vs. Practical people

Liveliness is an indication of how cautious or spontaneous you are.

Your culture has a lot to do with your level of liveliness, with people living in totalitarian societies being naturally more cautious and people living in liberal societies being naturally more spontaneous. In a similar way, your feelings about the authorities (police, politicians, officials) affect how cautious or spontaneous you are.

The same thing is true in your occupational culture. If you work for a high-tech development organization you're expected to have a high level of spontaneity. On the other hand, if you work for a security organization you're expected to have a high level of caution. If you work for a high-tech security development organization, you're expected to have high levels of both.

Whenever a dominant culture in a country feels another culture is a threat, changes in the level of liveliness occur. Extreme examples of over-caution occurred during World

Your Essence Interactions

War II when the German government labeled people of non-Aryan heritage as a threat and put them in concentration camps for execution. When the Japanese attacked Pearl Harbor, people of Japanese heritage who were living along the Pacific coast were rounded up and forcibly moved to special "War Relocation Camps". This happened even to United States citizens, including those who had been born in the United States and who had never been to Japan. Many other examples can be cited from different historical periods in countries around the world and you probably know several yourself. The suspicions of the dominant culture, even if unfounded, result in excessive caution for all cultures in the country.

Does your culture make it easy to travel? Then you have more flexibility to be spontaneous in deciding where and when you go. However, countries sometimes issue "travel advisories" for its citizens to be cautious when traveling to some places. No matter how cautious or how spontaneous you normally are, cultural differences in the places you go will affect how you act.

Abstractedness refers to your problem solving approach, whether you're more imaginative or more practical. French moralist Joseph Joubert said, "The man with imagination and no culture has wings without feet."

Throughout history there have been strong cultural influences on the individual's level of abstractedness. The great cultures of the world – whether they're Oriental, European, Slavic, African, Polynesian, or any other – have all had times when individuals were encouraged to be imaginative and other times when individual imagination was restricted.

New ideas threaten the status quo, so an innovative culture requires innovative leadership at the top. Cultures that encourage imagination and innovation produce great discoveries, inventions, art, music, and literature, but they also produce great thoughts for social change. Only strong cultures can encourage imaginative thoughts. Weak cultures, those in which the great majority of people have no influence on the leadership, must restrict imagination to survive.

Conscientiousness

Planned vs. Spontaneous Behavior

Task Performance Attitude refers to your work ethic, whether you're more careless or more obsessive.

Different cultures have different ways of measuring task performance attitude. The *Multidimensional Work Ethic Profile*, with 65 items that measure 7 different facets of work ethic, has been developed for Western cultures while a *Confucian Work Ethic*, which focuses on relationships and virtues, has been developed for Eastern cultures. While both have many of the same characteristics, they're directed towards different life styles. The main differences are on perceptions of hard work, self-reliance and centrality of work.

While strong individual personalities maintain the work ethic they want, culture exerts a strong influence. If your culture has assigned you to a role in life and there's little opportunity to either improve or lose your situation, you may easily become lackadaisical in your output quality. "After all," you argue, "it doesn't matter, no one cares, why should I bother?" This kind of attitude is reinforced when you receive substandard results from other people with the same attitude. It's a recurring problem that may eventually cause the culture to self-destruct. On the other hand, if you belong to a culture that rewards individual performance, you'll try to out-perform your associates to receive a larger reward. Interestingly, this "rewards based" culture also results in problems when the participants spend all their time working and little time enjoying the efforts of their labors. The saying, "Americans live to work while Europeans work to live" has some basis in truth.

Control Attitude concerns the manner in which you try to influence others, from being domineering to being submissive.

Cultural control has a significant effect on individuals in some cultures and almost none in others. Social clubs are cultures with expected control attitudes, but as voluntary organizations they're excluded from this discussion. After all, you knew what you were getting into when you joined and you can always quit. It's not so easy in other kinds of cultures. For example, the caste system in India permits members of higher castes to expect members of lower castes to exhibit a personality of deference towards them, and those lower caste members, in turn, expect to be treated as "servants" of the higher castes. This is similar to the command structure in the military. If you're incarcerated in a prison you quickly learn the control attitude others have over you and the options available to you. The hierarchical structure of administrator-doctor-nurse-patient in hospitals includes its own control attitude expectations. Some religions have certain activities and expectations for men and others for women, with women usually having subservient positions. Each different culture in your life includes its own control attitude.

Vigilance is the extent to which you trust others, from being suspicious to unconditional trust.

Cultures never trust one another. Countries have their spies in both friendly and unfriendly countries, and every once in a while we express shock (but are actually amused) by the capture of a friendly spy. Some social clubs have secret handshakes or passwords or rituals that only members of the club can know. Some stores have signs at the cash register that say, "In God we trust. Everyone else pays cash." Trust is a very personal attribute, and vigilance is the way to protect it.

Your Essence, Your Eternity

The United States government is organized on the principle of checks and balances, a primary type of vigilance. The three branches of government – Administrative, Legislative, and Judicial – are constantly reviewing the activities of the other branches to ensure that the American people can trust their government. Even so, independent polls frequently show a deep distrust of government. For example, a Gallup poll released in September 2013 reported that less than half of Americans trust their government to handle problems, an all-time low. If the people in a culture are suspicious of it, that culture can't expect trust from other cultures.

Some organizations include "team building" experiences as part of their efforts to build trust. This is done by forcing participants to rely on others to successfully complete an activity. For example, participants may have to climb over a wall, and the only way to do it is to get a boost from someone else. A more interesting exercise is to fall over backwards with the hopeful expectation of being caught by your partner. Circus performers must have unconditional trust in each other or catastrophes result. Some cultures can't exist without unconditional trust.

Your Essence Interactions

Extroversion

Level of Engagement with the External World

Privateness refers to the ease with which people can get to know you, whether you're more private or more open.

The culture in which you live has a lot to do with your level of privateness. Privacy is not a universal concept and was virtually unknown in some cultures until recently. In fact, many languages lack a specific word for "privacy" and have either borrowed a variation of the English word or use a phrase to describe it.

Some cultures have security personnel who collect information on people. In totalitarian cultures people learn from childhood that the more private you are the safer you are. Information is also collected in more open cultures, but it's not so obvious and not as threatening.

If you prefer to be more private, you're living in the wrong era. The advent of computer databases has made much of our previously private information available for others to see. Government regulations in some cultures require individuals to use specific identification numbers that link to much of their personal history. In addition to your financial history, your medical and sexual history is stored in databases that may be only marginally secure. If you find all this troubling, you're not alone. There's much continuing research on how to ensure individual privacy without compromising national security.

Material Attitude is the extent that material possessions are important to you, from being frugal to being wasteful.

For those of us brought up in highly materialistic cultures, it should come as no surprise that there's a scholarly periodical called the *Journal of Material Culture* and an entire academic discipline called *Cultural Materialism* that tries to understand why we need so many things. Basically, Cultural Materialism suggests that societies develop on a trial and error basis, producing only those things that the people in the society want and eliminating the rest. Cultural Materialism includes social structures such as law, government, religion, and family values, again keeping what people want and discarding the rest. From the perspective of **your essence** however, material attitude will focus on material possessions and discard the rest.

Your Essence Interactions

As a culture becomes more developed it becomes more material. As that bar is raised, however, a less affluent subculture remains excluded. As Mollie Orshansky noted:

> The wonders of science and technology applied to a generous endowment of natural resources have wrought a way of life our grandfathers never knew. Creature comforts once the hallmark of luxury have descended to the realm of the commonplace, and the marvels of modern industry find their way into the home of the American worker as well as that of his boss. Yet there is an underlying disquietude reflected in our current social literature, an uncomfortable realization that an expanding economy has not brought gains to all in equal measure. It is reflected in the preoccupation with counting the poor—do they number 30 million, 40 million, or 50 million?

But what is poverty in a culture? Perhaps Adam Smith said it best in 1776:

> By necessaries I understand not only the commodities which are indispensably necessary for the support of life, but what ever the customs of the country renders it indecent for creditable people, even the lowest order, to be without. A linen shirt, for example, is, strictly speaking, not a necessary of life. The Greeks and Romans lived, I suppose, very comfortably, though they had no linen. But in the present times, through the greater part of Europe, a

> creditable day-laborer would be ashamed to appear in public without a linen shirt, the want of which would be supposed to denote that disgraceful degree of poverty which, it is presumed, nobody can well fall into, without extreme bad conduct. Custom, in the same manner, has rendered leather shoes a necessary of life in England.

Cultures generally posit that excessive materialism is bad but the more affluent ones continue to produce as many different things as possible to expand their economies. Ger & Belk studied the cultures in Romania, Turkey, United States, and Western Europe and found both justifications and excuses for overt materialism. The justifications were that people needed the latest gadgets, that people wanted to feel that they owned something special, and that people thought their spending was essential to the welfare of the country. The excuses were that society expected them to buy certain items, that everyone else was getting them, and that they deserved a few luxuries. They also found that these arguments were pervasive and morally justified by both the individuals and the cultures, in spite of continuing condemnation of excessive materialism.

Risk Attitude is the extent of your willingness to take chances, from being guarded to being reckless.

The most significant cultural influence on risk attitude is the degree with which the culture permits equality and representation for its members, as noted by Weber & Hsee. People in cultures with less equality are naturally more risk adverse; people in cultures with more equality are naturally more risk adventurous. The culture can be your government, your employer, or your social organizations. If the culture in one area (such as a social club) permits more equality then you're naturally more risk adventurous there, even if other areas (such as your government) are more restrictive and cause you to be more risk adverse when dealing with them.

Agreeableness

Compassion vs. Antagonism

Ethics is the extent to which you follow the right action for the greater good, from self-sacrifice to self-gratification.

Ethics is much broader than the difference between right and wrong, signifying actions that are both personally satisfying and that result in positive outcomes for others.

Your Essence, Your Eternity

To provide a clear understanding of what constitutes ethical and unethical behavior, cultural laws have codified many situations. "Thou shalt not murder", "Thou shalt not steal", and "Thou shalt not bear false witness" are universal across cultures, although some others of the Ten Commandments are culture-specific. As cultural anthropologist Ruth Benedict noted about an individual's view of the world:

> His very concepts of the true and the false will still have reference to his particular traditional customs. (p. 2)

Some established cultural aspects, such as slavery and restrictions on voting rights, have been reconsidered and re-established as unethical and illegal in many cultures. But don't think that societies have become enlightened and are finally correcting past unethical behaviors. All this has happened before. For example, the biblical Book of Exodus describes how Egypt freed Israelite slaves around 1440 BCE, and Cyrus the Great of Persia abolished slavery around 539 BCE. Greece and Rome established slavery when they controlled much of the Western world, but classical slavery pretty much disappeared in the West during the Middle Ages (although debt slavery was common). Back again during the 15th through 18th centuries, slavery was abolished in Western Europe and North America in the 19th century. What's considered ethical at one time can be unethical later, but become ethical again.

What do you consider to be ethical behavior? Is euthanasia ethical? What about capital punishment? How about abortions? Should prostitution be illegal? Do animals have the right to expect ethical behavior from us? There are passionate arguments on both sides of these questions, and others, which help to define our culture.

Our culture, in turn, helps to shape our essence. We see what our culture feels is ethical and unethical, and those feelings have a large influence on our personality. If you move from one culture to another with different rules for ethical behavior, you'll probably think your new culture is wrong. If you eventually accept the new ethics, your personality will have changed and your essence will be affected.

Morals are the extent to which you respect the principles of society, from being self-indulgent to being rule-bound.

"Respect" is different from "accept". Individuals who both respect and accept the principles of society are its model citizens. Individuals who respect the principles of society without accepting them become social activists intent on changing those principles to ones they can accept. Individuals who accept the principles of society without respecting them are constant complainers who try to "work" the system to find loopholes to their personal advantage. And individuals who neither respect nor accept the principles of society are its criminals.

Your morals are your sense of right and wrong, your personal code of conduct. The collective senses of right and wrong for everyone in a culture define the morals of that culture. As with ethics, those morals are codified into cultural laws that all must obey. Different cultures may have different morals that are reflected in different, and sometimes opposite, laws. For example, smoking marijuana is illegal in some countries and legal in others. These cultures each think they're right and the other is wrong. From the perspective of your essence, both are right. Haidt & Joseph investigated "why morality varies so much across cultures yet still shows so many similarities and recurrent themes" and proposed five psychological traits:

Your Essence Interactions

1. **Harm/Care**, in which we feel the pain of others and leads to kindness, gentleness, and nurturance
2. **Fairness/Reciprocity**, which generates our sense of justice
3. **Ingroup/Loyalty**, in which tribalism leads to feelings of patriotism
4. **Authority/Respect**, which includes our feelings for authority and tradition
5. **Purity/Sanctity**, in which our feelings of disgust generate our sense of bodily purity

In general, different cultures place more emphasis on some of these traits and less on others. For example, more liberal cultures emphasize fairness/reciprocity and harm/care over the others, while more conservative cultures are the opposite. Other moralists and psychologists agree or disagree with these classifications, so you can freely decide if they're sensible, and you can find distinguished researchers to back you up.

Of course, not everyone within a culture has the same sense of right and wrong, which leads to claims of discrimination when rules are enacted that favor (or disfavor) one group over another. To avoid continuous civil wars, many cultures have enacted a challenging system in which individuals who feel they've been wronged can petition the culture, through legal representatives, to change the rules. While this challenging system has been abused by frivolous challenges and outrageous demands by greedy individuals, the overall effect on society and on the culture has been positive. To improve this system it should not be necessary, as Shakespeare suggested in Henry VI, "The first thing we do, let's kill all the lawyers."

It's clear, then, that the development of a culture is greatly influenced by the morals of its individuals, and the morals of each individual are greatly influenced by the moral rules the individual experiences within the culture.

Sensitivity is a measure of how tough-minded or tender-minded you are, from being completely impersonal to being extremely sympathetic.

Sensitivity is so strongly influenced by the culture that individuals born with the "wrong" level of sensitivity may have a difficult life. For example, a person who believes that rewards should be based on individual initiative (a capitalistic perspective) but who's living in a culture that strongly supports caring for all individuals (a socialistic society) will be seen as insensitive and even mean-spirited. That same individual living in a culture controlled by a strong-willed ruler (a totalitarian society) will be seen as overly sympathetic and, perhaps, as a threat. Rule #1 in a totalitarian society is not to be seen by the leader as a threat.

Of course, there are sub-cultures within every culture, and while you may be born into a culture with a strong internal sensitivity, your sub-culture is your own choice. We see examples of extremely greedy people who only seem to want to increase their own wealth, often at the expense of others. We also see examples of extremely caring people who seem to sacrifice their personal comfort to help others who are less fortunate. Both groups look at the other as if they live on a different planet. These sub-cultures exist now, have existed throughout history, and will exist in the future because there are always new individuals being born with those interests.

Aggressiveness is the manner of your interaction with others, from being predatory to being loving.

From a cultural perspective there's a distinct difference between aggressiveness towards individuals within your culture and towards those in other cultures. The culture can be your family, your country, your religion, your ethnic group, whatever you consider to be related to you or not related to you. The aggressiveness can be physical, emotional, financial, or any way the individual can exhibit their superiority. But however it's done, as Supreme Court Justice Potter Stewart is said to have remarked about pornography, "I know it when I see it."

Within a culture aggressiveness is used to show dominance. With many animals this dominance is to establish a breeding hierarchy. Humans use intra-culture aggressiveness for more purposes, although watching teenage boys compete for girls is not so different from watching any male in any species compete for the females of their choice. The gender of the participants seems to make a difference, too, with male-male and female-female encounters being distinctly more aggressive than male-female encounters.

Some cultures have been described as very low on aggression, such as the Kung Bushmen, described by Elizabeth Thomas as based on their lack of possessions to fight over. But other factors, such as status and mating opportunities, would raise the aggression score for these people. Sometimes athletic events between cultures stimulate aggressive behavior in both cultures.

Other cultures, such as the natives of Bellona Island in the South Pacific, are based on violence caused by feuds lasting for hundreds of years. Rolf Kuschel has recorded and analyzed the oral narratives of these events to try to understand why they like to kill each other. Since there were never more than about 450 people on the island, having a culture based on extended violent feuding is a real threat to cultural survival.

Leslie Sponsel suggests several factors that foster peacefulness in cultures:

- The beliefs and religious convictions they have strengthen their daily nonviolent lives.
- They have sanctions against aggressive behavior.
- The societies are often isolated or remote.
- Flight rather than fight approach when threatened.

Fairness is the extent of recognizing other points of view, from being consciously biased to being consciously impartial. Fairness related to your culture is both internal and external – how fair the culture is with its members and how fair it is with other cultures.

Financial fairness is a good barometer of overall fairness because money is always a sensitive subject. The worldwide financial downturn starting in 2007 highlighted the difficulties faced by some cultures. For example, Greece suffered a near collapse of its economy in 2010 partly because of corruption in tax collecting. One comment was that Greeks were all in favor of higher taxes as long as other people paid them. When the people see their culture as inherently biased in some way, the people themselves assume that bias. By trying to remain fair while everyone else is biased, you put yourself at a distinct disadvantage in cultural activities.

We've all seen news stories describing how unfairly one culture is being treated by another, a bias that often has a long and continuing history. The result of this bias is an innate suspicion about anything the other culture says or does, further complicating attempts to improve the relationship. This distrust is compounded by the leaders of the cultures, who sometimes became the leaders by inflaming the passions of their members against the other culture. Outsiders who try to calm the situation are either ignored or mollified until they go away. Cultural unfairness is a long-term problem that requires exceptional leaders on both sides to correct.

Now focus cultural fairness down to your personal opinions. We all belong to cultures of one kind or another and we all deal with issues from other cultures. Inspect your own beliefs to detect biases not based on facts but based on how your culture feels about the other. Some of these biases may be unfairly positive, especially if another culture disagrees with a culture you don't like either. This follows "The enemy of my enemy is my friend" philosophy. The more you can elevate yourself from this quagmire of biases to a position of impartial analysis, the fairer your own personality will become.

Neuroticism

Tendency to Experience Negative Emotions

Tension concerns the level of your patience, from being comatose to being demanding.

Cultures around the world are very different with regard to the patience expected of people living there. If you belong to a culture that expects you to wait on long lines, you either develop that level of patience or you do without. If you visit another culture, don't import yours. For example, in Russia tourists are valued and permitted to go to the front of store lines. A Russian tourist visiting the United States saw the long lines at a supermarket and announced, "tourist, tourist" and cut in front. You don't want to find out about cultural differences by an angry mob.

Your Essence, Your Eternity

When people in the United States go to a restaurant, they expect and generally get rapid service; when people in France go to a restaurant, they expect and generally get a lingering dining experience. When French people come to the U.S. they can't understand why Americans are in such a rush; when Americans go to France they can't believe how long it takes for things to happen. To minimize the amount of tranquilizers you take when traveling to another culture, follow the old saying "When in Rome, do as the Romans do."

Other parts of your life have their own cultural tensions. If you choose a career such as a stockbroker, air traffic controller, or surgeon your workday is very fast paced. If, instead, you're a teacher, farmer, or business analyst your workday, while demanding, is paced at a much slower rate. If you're a policeman, firefighter, or soldier your workday includes long periods of quiet time interrupted by short bursts of intense pressure. Your choice. Many people pursue a career without realizing the demands associated with that work and find their lives surprisingly unhappy. A little patience while choosing your life's goals will be time well spent.

Essence Interaction with the Habitat

Habitat is always important to all living things except, curiously, humans. One of the traits of humans is our ability to profoundly change our habitat to better suit our needs, and we've done that since the earliest humans came to be. The more "civilized" a culture becomes the more that culture is separated from its habitat. Warber & Irvine describe how humans have become sequestered within their self-made "inside" environments rather than in the natural world. In most large cities the nighttime sky is filled with artificial lighting rather than the panoply of stars seen in the country, and children are raised with little understanding of life outside the city.

Malidoma Somé describes a different view of the habitat in Africa, where natives see spirits in every object and where strong connections to the *spirit world* are natural and frequent occurrences. The same perspective is felt in native cultures in other parts of the world.

While most of your interaction with your habitat affects your physical body, there are also habitat interactions with your essence. Here are some of the important ones.

Openness

Imaginative vs. Practical people

Liveliness is an indication of how cautious or spontaneous you are.

Your culture affects your level of liveliness in some ways and your habitat affects it in others. Do you live in an environment with many older people? Then you're probably more cautious. On the other hand, if you live in an environment with many children, you're probably more spontaneous. Spontaneous people living in a cautious habitat are considered reckless; cautious people living in a spontaneous habitat are considered boring. Your level of liveliness as compared with your neighbors is a distinct factor in whether you stay there or move.

Your level of caution or spontaneity is also affected by how safe you think your habitat is. If you live in a high-crime neighborhood you'll be more cautious whenever you're outside and you'll certainly keep your possessions secured. On the other hand, if you live in a gated community with friendly neighbors you know well, you're probably less cautious. People who move from a small, friendly town to a big city often find themselves in some peril because their natural trusting nature from the country becomes a green light for abuse in the city. Familiarity with your habitat is important for your survival.

Abstractedness is about how you solve problems, with more imagination or with a more practical approach.

To understand your abstractedness level with regard to habitat, it's your habitat at the moment that should be considered. When you're in a familiar habitat you're usually practical because you know what has worked and what hasn't, and you want to do what works. When you're in an unfamiliar habitat you're usually imaginative because you don't know what to do, so you try to figure it out.

For example, surgeons follow established procedures that have been refined over many years to best ensure the success of the operation and the safety of the patient. Those procedures are established for the "normal" operation. At the same time, the surgeon must be ready to innovate if s/he finds an unusual situation and has now been placed in an unfamiliar habitat. Doctors practice the art of medicine, with "practice" noting its practical aspect and "art" noting its imaginative aspect. If you think about it, activities as diverse as mountain climbing and playing chess demand both practicality and imagination depending on the kind of immediate habitat you find yourself in.

Some organizations (or people) want to restrict the way activities are done to a single way that they consider the "right" way. The trouble is that most people hate to be boxed in by single methods and look for places that give them more freedom. To get around this negative perception the concept of *best practices* has evolved. The idea is to consider all the ways to achieve a goal and then pick those that achieve it most efficiently. There are even awards and recognition for developing "best practices" and individuals feel superior when they use them. However, if those "best practices" are found to be inferior but are still required, the façade has been lifted and the participants know they've been tricked into following arbitrary rules. The point is, if you're going to restrict someone's imagination, make sure your way is the best.

You can also be involved in different familiar habitats that require different levels of abstractedness. Let's say you work on a factory assembly line attaching widgets to gizmos during the day and you write science fiction stories at night. Assembly line work is very coordinated, with each person expected to do his/her job just as they're supposed to, all the time they're there. This is an extreme example of required practical abstractedness because any innovation in the process will change things and likely slow down the assembly. On the other hand, your science fiction writing requires out-of-the-box imagination because readers of science fiction demand original thinking in the stories. For you to be successful at both occupations, you have to adjust your abstractedness level depending on the habitat you're in at that time.

Extroversion

Level of Engagement with the External World

Material Attitude is the extent that material possessions are important to you, from being frugal to being wasteful.

Do you have bulging closets or overflowing cabinets? You may have too many possessions for your habitat. There's a well-known belief that possessions are acquired to fill the space available for them. Sometimes, though, desire overwhelms need and more and newer and bigger and

Your Essence, Your Eternity

brighter things are bought until you have more possessions than you can possibly use. In desperation you gather the less important items and give them away. Unfortunately, for people with a wasteful material attitude, this is only part of the cycle that continues with finding new items to buy.

Among frazzled workers going to and from a job they basically dislike is a feeling that the whole material acquisition concept is a *rat race* they'd like to avoid. The Hadza, people living around Lake Eyasi in northern Tanzania, seem to have succeeded at this avoidance. Michael Finkel, in his studies of this pre-agriculture society, found them living a hunter-gatherer existence similar to the way all peoples lived 10,000 years ago. They grow no food, raise no livestock, and are free from schedules, jobs, bosses, laws, news, calendars, money, and worry. Finkel admitted that he envied this freedom, but he did return to the United States to write his report.

If you have a reasonable material attitude your habitat is one control for your possessions. Over time your habitat changes. A young person living with his/her parents has a rather limited space for possessions. After leaving home that space expands and continues through middle life until old age, when people usually want to live in a smaller place that's easier to maintain. That's often a difficult time because many possessions that have been so necessary for the majority of life now must be discarded and only a limited set of special possessions are kept. Advanced possession planning will make each stage of your life easier.

Risk Attitude is your willingness to take chances, from being guarded to being reckless.

Your risk attitude has a lot to do with your habitat. For example, field mice spend a lot less time in the field if they can find adequate food around a human dwelling because of the high predation rate (hawks, etc.) in the field. Meerkats, foraging animals in the Kalahari Desert, assign one animal as a "sentry" to watch for predators while the rest of the group searches for food. The domestication of the dog has resulted in its evolved inability to live in the wild. As Karen Lange quotes biologist James Serpell, "The domestic dog exists precariously in the no-man's land between the human and nonhuman ... neither person nor beast." Each species has developed its own approach towards risks in its habitat.

Some habitats have their own innate physical risks. Some places in the world are prone to tornados, earthquakes, hurricanes, drought, flooding, avalanches, landslides, or temperature extremes. Any thing living in those habitats has to understand their environment and make a suitable accommodation or suffer the consequences. Sand crabs on beaches have to dig out their dwelling after being flooded by every incoming wave. People living on those same beaches have to rebuild their houses when they're wrecked by frequent storms. Both stay in their habitats but, to be fair, the sand crabs have nowhere else to go.

Habitats change over time and the risks associated with them change accordingly. If the risks become too great (lack of food, temperature out of range, loss of habitat) then the things living in that habitat must either find a more suitable habitat or perish. We've learned about previous habitat catastrophes (asteroid impacts or huge volcanic eruptions) that have caused such worldwide climate changes that there was nowhere to go, resulting in mass extinctions.

Human society has created habitats with their own inherent risks. People caught up in wars or environments controlled by criminals find themselves having to take extraordinary measures just to survive. Businesses with sensitive information or the need for access control require their employees to work in restrictive habitats. When you're traveling it's red light stop, green light go. You must know the risks and adhere to the constraints of your habitat to be successful in it.

Socialization is the extent you fit in with a group, from being anti-social to being actively social.

The degree of socialization in a habitat depends on how high or low the living thing is on the food chain. Small or relatively weak animals tend to band together for their overall safety. Although the rhinoceros eats plants and the lion eats animals, rhinos aren't particularly afraid of lions. Rather the other way around. Size and armor matter. Back in the good old days, when humans and lions and rhinos were competing with each other in the same habitat, those humans who didn't socialize together became meals for other, more physically capable animals.

Reptiles, for example, spend much of their time trying not to be eaten, so they're very particular about their habitat and are nervous when placed in an unfamiliar one. That's why people who buy a snake as a pet should be especially careful at first because the snake may strike at anything that moves in its new environment. Besides, cold-blooded creatures such as reptiles prefer hot, humid climates, probably quite different from your nice comfortable house.

Your Essence, Your Eternity

A male mountain gorilla, at 450 pounds of mostly muscle, would be comfortable in your house, although you may not be. Each living thing has its habitat of preference and its social contacts of preference.

Do you think the more people available for socialization, the greater the likelihood of socialization? In other words, are people more social in big cities or in small towns? Controversy exists, with everyone saying they're more social than the other place. However, there are plenty of urban legends that urban areas are more lonely. A report in 2008 by the New York City Department of City Planning found that one of every two apartments in Manhattan had only one occupant, by far the most solitary living arrangement in the United States. Perhaps you'd rather believe Mark Twain, who called New York "a splendid desert – a domed and steepled solitude, where the stranger is lonely in the midst of a million of his race." As desirable as cities are for employment, cultural experiences, and educational opportunities, few people would call them particularly social.

City dwellers who yearn for the kind of friendly socialization they see in movies about small towns think they can just find a gorgeous picturesque village, move there, and instantly be part of the community. What this romantic dream fails to recognize is that small towns have ingrained relationships, making it difficult for newcomers to slide in. In addition, people often stay in small towns to escape the problems they see in movies about big cities, and they're understandably concerned that someone from a city may bring those problems (or attitudes) with them. Newcomers should see their new village as a new culture and a new habitat that must be learned by them. Socialization is certainly possible if you're patient and willing to spend the necessary time to develop it.

The Internet is a special type of habitat, with some people spending so much time online that they practically live there. There are two types of Internet activities that relate to socialization: online games and social networks.

Online games involve players at different locations who choose an identity in the game and then participate with and/or against other players. The identities are called *avatars* because they're extended representations of the player, often with special abilities (strength, invisibility, etc.). Some of these games have the players in an artificial habitat that may or may not look like Earth. Players sometimes become so involved with their online personalities that their real personalities become secondary and they start to believe they're the avatar. Losing touch with reality is never a good socialization technique.

The social networks are incredibly popular Web sites used by people all over the world to make new friends and develop new relationships. Interestingly, they were created by the biggest social outcasts on the planet – those technology fanatics who spend their lives writing computer code in part because they relate to computers much better than they relate to people. The social networks are rather addicting, with some people spending much more time typing messages to other addicts than they spend talking face-to-face (or even face-to-phone) with people. The situation can reach the absurd when people in the same room communicate via typed messages rather than talking to each other. The unfortunate and predictable result is a reduction in personal communication skills, making it more desirable for those people to continue writing messages because they're forgetting (or even forgotten) how to relate face-to-face.

The solution to excessive Internet use is the same one used by parents who want their children to stop watching TV and go out and play – pull the plug. You can't improve your people socialization skills if you don't socialize with people.

Neuroticism

Tendency to Experience Negative Emotions

Emotional stability is a measure of your patience when frustrated by difficulties, from being determined to panicking.

When your habitat is mild, being patient is easy. The true measure of your emotional stability is when your habitat is a challenge.

Habitats may be challenging all the time or they may become challenging. When you're regularly challenged your emotional stability is either raised to a higher level or you lead a stressed life. Such habitats may be physical (i.e. frequent storms) or emotional (i.e. lifestyle). Individuals choosing careers such as air traffic controller, bull rider, or stock exchange trader must always be totally focused to succeed. Taking the time to assess your innate emotional stability relative to your possible career choices may save you a lot of difficulties in the future.

Your Essence, Your Eternity

Quiet habitats that change are the most stressful because the individuals in them usually aren't prepared for the change. Only people with a relatively calm level of emotional stability can succeed when unexpected habitat changes occur.

By understanding your essence traits and their interactions with your culture and habitat you see the reasons you are who you are. How did you get that way? By the different ways you've communicated with your physical body, culture, habitat, and yes, even with your essence. But a lot of what happens to you just seems to happen. Why? Here's why.

Chapter 11 – Your Life of Chance

> As a matter of fact, I believe that success is 95 percent luck and 5 percent ability. Take my own case. I know that there are any number of men in my employ who could run my business just as well as I can. They didn't get the breaks – that's the only difference between them and me.
>
> Julius Rosenwald
> Past President of Sears, Roebuck and Company

Getting the Breaks

When someone "gets the breaks" they call it good luck. When someone doesn't "get the breaks" they call it bad luck. Neither exists. Good luck is a chance event that goes in our favor; bad luck is a chance event that goes against us. The chances for all possibilities exist and can happen according to their respective probabilities. There's no such thing as "luck".

Just because luck doesn't exist doesn't mean a person won't believe in it. Belief in good luck has existed since the first caveman sold the second caveman a good luck charm. Similarly, bad luck has existed since that second caveman, after finding out his luck didn't improve, put a curse on the first caveman.

Moving forward, we see strong relationships between ritual and luck in both ancient and modern religious practices. Mesoamerican religions, including the Aztecs, Mayans, and Incas, used their rituals to appease their spirits. Voodoo and similar magic-oriented religions continue those traditions, albeit without the human sacrifices seen in earlier cultures.

Judaism, Christianity, and Islam don't believe in luck, expressed in the Bible as:

> The race is not to the swift or the battle to the strong, nor does food come to the wise or wealth to the brilliant or favor to the learned; but time and chance happen to them all. (Ecclesiastes 9:11)

Guru Nanak Dev, founder of Sikhism, is even more direct:

> Luck is but lack of self-confidence and fruit of idleness.

However, belief in luck is a strong part of most religions anyway, even when the founders preached against it. Buddha taught his followers to believe in moral causality and not to believe in luck, with less success than he had hoped. In Thailand, for example, Buddhists wear lucky amulets that have been blessed by monks, no less. Traditional Indian, Japanese, and Chinese religions include lucky gods. Hinduism has a deity named *Lakshmi* for money and fortune who, if you follow a meticulous prayer procedure in Sanskrit, will give you his blessing. Believers of Wicca include luck in their practices, using spells and magic to influence the outcome of events. Many followers of most religions believe that certain rituals must be followed to ward off evil spirits. If you have a problem that's caused by a spirit, no matter what your religion, there's an incantation to minimize it.

Figure 11-1 is an advertisement to wear the Sun Stone for "the sun's mysterious power for healing, wealth, and happiness" with "strange tales abound of lucky strikes, sudden good fortune in love, in games and various undertakings."

Your Essence, Your Eternity

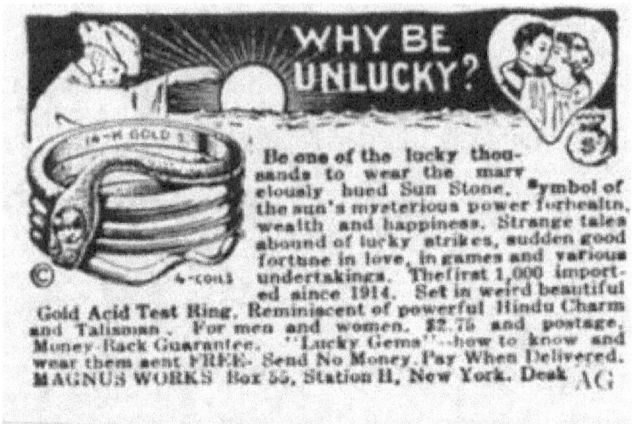

**Figure 11-1. Advertisement for Lucky Jewelry
From 1926 Art and Beauty Magazine**

There are also lucky symbols. In general, horseshoes are considered lucky (to give you luck when entering a room one is nailed above the entrance – your best luck will be that it doesn't fall on your head!), as are wishing wells, wishbones, crossed fingers, four-leaf clovers, and spotting a shooting star. Finding a pot of gold is lucky by default. Here are some lucky symbols in various countries:

> China – jade, Buddha figurines, and the number 8
> Egypt – scarabs
> France – garden Dwarves
> Germany – pigs and ladybugs
> Ireland – representations of Leprechauns
> Italy – fish
> Russia – carrying a fish scale in your wallet or purse
> Scandinavia – representations of Trolls
> United States – rabbit's foot and the number 7

Your Life of Chance

Lucky charms may actually improve your performance. Stuart Vyse found that volunteers who carried lucky charms performed better in games than those volunteers whose charms were confiscated. Follow-up tests confirmed that the talismans increased the player's confidence by giving them the illusion of control.

Feeling lucky? There's a game of chance for you. Consider the habitat you'll find in a typical casino. Drinks are often free because, even though the house has a probability advantage over the players, that advantage gets larger the fuzzier your mind becomes. There are no windows or clocks so it's easy to lose track of time. The slot machines use a coin made of a material that makes a lot of noise when it hits other coins, making the winnings sound larger. Just in case you're a little hard of hearing, bells go off, too. And in case you're deaf, lights flash. Everyone seems to be winning all the time! Get me to a table!

One of the easiest games to play is roulette, but the easier a game is to learn the less chance you have of winning. For roulette, the house has a 5.26% edge over you. You'll get better odds playing the slot machines, but you'll still lose overall because the best ones pay back 98% of what you put in, and there are very few frequent-payoff machines. The game with the best odds is blackjack, but the house still has a small advantage over you. The point is that all games of chance are based on odds (probability), not luck. Sure, occasionally someone "beats" the odds by being one of the small group on the probability chart that does win, and that person comes away with a jackpot. The casinos are quick to congratulate winners because they want everyone to think they can do the same thing. Keep in mind that that expensive building with all the fancy furniture and fancy barmaids wasn't built by people who expect to lose money.

If you don't go to casinos but still want to get rich quick, many governments sponsor lotteries. For only a small investment you can win a fortune. But returning to basic principles, the government isn't running the lottery to make you rich. They're running the lottery to make the government rich.

Let's look at an example. Let's say you have to pick 6 numbers out of 49 possible numbers to win the jackpot. The odds of doing this are 1-in-13,983,816. If you buy 10 different groups of numbers (and spend 10 times as much), your odds improve to 1-in-1,398,381. That's still over a million-to-one odds against you and you've just spent the

Your Life of Chance

kid's lunch money. It's been said that you have a better chance of dying on the way to buy a lottery ticket than you have of winning the lottery. Not a very good investment.

But people buy lottery tickets all the time because:

1. Somebody wins. You see their smiling picture every week. Somebody wins because so many people are playing that somebody will be the 1-in-14 million, but it still makes you jealous.
2. It's run by the government, so you're basically paying yourself, right?
3. Everyone can afford to buy a lottery ticket.

There are also people like Evelyn Adams who won the New Jersey lottery twice, for $4,000,000 and then for $1,500,000. And Virginia Fike bought two Virginia lottery tickets one day and matched five of the six numbers on each one, winning $1,000,000 twice. If they can do it, why can't you? As the lottery ads say, you can't win if you don't play. Of course, you could also wind up like Maureen Wilcox, who bought tickets for both the Massachusetts and Rhode Island lotteries. She picked the winning numbers for both, but the numbers for Massachusetts won in Rhode Island and the ones for Rhode Island won in Massachusetts.

So, if you play the lottery or any other game of chance, just remember the odds you have of winning rather than the luck you think you'll have to win.

Probabilities

Luck has no effect on your essence, but chance events do. Everything you do has a relative probability for each outcome, for the success you're attempting, for the failure you may receive, and for everything in between. Here are the chances of a few specific events happening to you, although you may see different numbers from different sources:

- **Finding a four-leaf clover** – 1 in 10,000, so getting lucky requires a lot of luck on its own.
- **Becoming a billionaire** – After you find that four-leaf clover, you still have only a 1 in 1,000,000 chance of becoming a billionaire if you live in the United States, 1 in 7,000,000 if you live elsewhere.
- **Being audited by the IRS** – 1 in 100 each year, but if you earn over $100,000/year it increases to 1 in 60, so becoming a billionaire does have a bit of a disadvantage. If you've previously been audited and you've had to pay a penalty, your chances are much greater of being audited again. Don't steal – the government hates competition.
- **Being struck by lightning** – 1 in 500,000 each year but only 1 in 2,500,000 people will be killed by a

strike. If you stand under a tree or walk on the beach during a lightning storm your chances are much higher and you probably won't be reading this.

How do these chances happen, and is there anything we can do to change them? The answers to these questions come from probability theory and we'll explain a little of that now, but don't worry – no math will be involved. We'll use pictures to explain the concepts. Everyone likes pictures.

Activities in which the most likely possibility is the average possibility follow a *normal curve*. All honest games of chance, including lotteries, follow a normal curve. A typical normal curve is shown in Figure 11-2. As you can see, the most events cluster towards the middle, which represents the average, with less likely events shown towards either end. By convention, a vertical line drawn through the peak is considered the 0 (zero) position, with negative positions shown to the left of that line and positive positions shown to the right. A typical lottery ticket buyer, such as you, will get the average reward, namely 0. People like Evelyn Adams are also on the normal curve, but far to the right. People who play the lottery every day and never win anything are on the normal curve, too, but far to the left.

The normal curve accommodates everyone involved in an activity that clusters towards the middle.

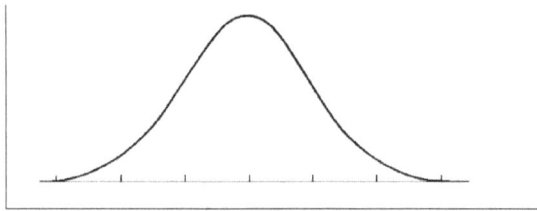

Figure 11-2. Typical Normal Curve

The amount of submissive or aggressive chemicals secreted by a physical body follows a normal curve, with most physical bodies producing the normal amount of both, but (as the normal curve shows) some physical bodies will produce more (and some much more) or less (for some, much less) of a chemical, affecting the natural behavior of that individual. The distribution of personality types also follow a normal curve, as described in Chapter 9 under *Essence Maturation*. When both the amount of body chemicals and the personality type fall in the same section of the normal curve, you feel comfortable in your life. When the area of the normal curve for chemicals is different from that for personality type, in other words your physical body produces more aggressive (or less submissive) chemicals and your personality type is innately more passive, a conflict arises. A similar conflict will arise for physical bodies that produce more submissive (or less aggressive) chemicals and your personality type is innately more aggressive. These situations are described more fully in Chapter 9 under *Personality vs. Behavior*.

There are also distributions that follow a *skewed* curve. This happens when something affects the distribution to force it one way or the other, and several examples are shown below. Figure 11-3 shows what happens when a distribution is skewed to the left, with most of the events on the left side of the zero line. Figure 11-4 shows what happens when a distribution is skewed to the right, with most of the events on the right side of the zero line. And Figure 11-5 shows what happens when there are two equally likely outcomes.

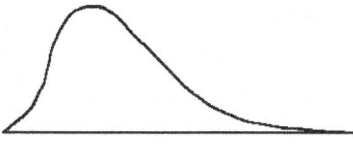

Figure 11-3. Most events occur to the left of center

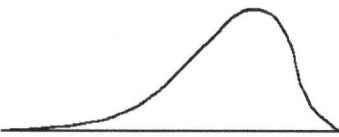

Figure 11-4. Most events occur to the right of center

Your Essence, Your Eternity

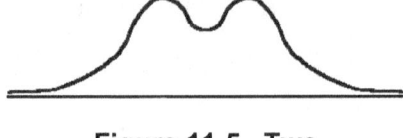

Figure 11-5. Two equally likely outcomes

What can cause such effects on a normal distribution? There are three possibilities:

1. You can purposely skew a result by taking actions that improve the chances of one event happening over another.
2. Explainable actions can happen without your knowledge that skew the results.
3. Unexplainable actions can happen that skew the results.

For the first possibility, here are examples where you take actions to skew the results. If you invest in a university education your lifetime income will probably be skewed higher than average. If you work as a sparring partner to a professional boxer, you're more likely to suffer injuries than the average person. If you're shooting craps and are using loaded dice, you're more likely to win than the average player. You're also more likely to suffer injuries, but that's the chance you're taking. For the bipolar case, you may have two gorgeous girlfriends that you date more frequently than anyone else. Of course, this example may become similar to the injury example if one of the girls finds out.

Sometimes you think you're skewing your chances, but you're not. For example, let's say three men, Larry, Curly, and Moe, decide to get rich quick by investing in the stock market. Larry takes stock investment courses at the university, Curly hires a professional stock analyst, and Moe buys a dartboard. Who has the best chance of making money? Alessio Biondo et al., say it's Moe. Their analysis of 15-20 years of historical data from the major stock exchanges found that a computer program choosing stocks at random did as well as professional financial consulting, and was much less risky. Try not to dwell on what that says about the people who manage your company's retirement fund.

For the second possibility, here's an example of an explainable but unexpected action that has taken place and skewed the results:

During the Vietnam War the United States Selective Service System instituted a draft lottery to select new soldiers. The first drawing was held in 1969 and was designed as follows.

Your Essence, Your Eternity

The days of the year (including February 29) were written on slips of paper as numbers from 1-366 and were put individually into separate plastic capsules, placed in a shoebox, and then dumped into a deep glass jar. The capsules were drawn one at a time on national television by dignitaries with great fanfare. It soon became apparent that people born late in the year had a greater than expected chance to have been picked sooner, i.e., to be drafted before others. After some investigation it was revealed that the capsules weren't mixed well before the drawing and, since they were filled starting with 1 and ending with 366, those born later had capsules towards the top of the jar. Irrespective of the howling from young men born in December, the results were permitted to stand but later lotteries were more carefully monitored.

The third possibility, in which unexplainable actions happen that skew the results, is more complicated. It could be that several unlikely events have occurred together to result in an unexpected outcome. Each of the events has a small probability of happening, with a smaller probability of happening together, but that probability is not zero. Such a situation, if it's not too strange, is brushed off as a coincidence. Even if it's very strange, it should be considered just one of those crazy things that happen sometimes.

For example, the author Anne Parrish and her husband were browsing in a bookstore in Paris in the 1920s when she came upon a special children's book. It was a well-worn edition of *Jack Frost and Other Stories*. She mentioned to her husband that the story had been one of her favorites as a little girl. He opened the book and was stunned to read the inscription inside: "Anne Parrish, 209 N. Weber Street, Colorado Springs, Colorado."

In another example, in 2009 retired judge Edwin Collier and his wife were hit and badly injured by a drunken driver. This was the same man Judge Collier didn't send to jail in 1998 for drunken driving, even though he had also been arrested for drunken driving just 3 months before. To be fair to Judge Collier, the sentence he imposed (supervised probation and abstinence from drinking) was a standard sentence at the time.

Also in 2009, Johanna Ganthaler, a pensioner from Austria vacationing in Brazil, arrived late for her flight home on a plane that later crashed into the Atlantic, killing everyone on board.

Your Essence, Your Eternity

Unfortunately, there are people promoting their own agendas who will claim that tragic event that happened to Collier and the fortunate event that happened to Ganthaler were acts of judgment from a Supreme Being rather than the coincidences they really were. People who claim such things have their own tragedy – a miserable essence – and should spend their time improving themselves rather than criticizing others.

Sometimes, though, very strange and unexplainable things happen in your life and you wonder about them. You can probably think of a couple of times it seemed that someone was watching over you and saved you from a potential disaster. This concept was employed by ancient Greek playwrights such as Horace and Euripides and is called *deus ex machina*, which means "god of the machine." Essentially, the plots of some plays became so complicated that the only way the hero/heroine could survive was to have a god appear (lowered by a crane or rising up through a trap door) to save them. Sound like your situations?

Here's a real-life example. In the 1930s a street sweeper named Joseph Figlock was cleaning an alley when a baby fell from a fourth-story window and landed on him. Figlock and the baby were injured but recovered. The next year Figlock was cleaning another alley and a 2-year-old child fell from another fourth-story window on him, with the same results.

The most striking examples happen when a person is under a lot of stress. When Ernest Shackleton's ship became frozen in Antarctica in 1916, he and two sailors went for help. He said, "I know that during that long and racking march of thirty-six hours over the unnamed mountains and glaciers, it seemed to me often that we were four, not three."

Biologists suggest that such things are hallucinations caused by sleep deprivation, oxygen deprivation, low blood glucose, exhaustion, monotony, or other physical stresses that confuse the brain. Peter Suedfeld found that these phenomena sometimes offer useful information or advice. Griffith Pugh, the physiologist on the 1953 conquest of Mount Everest by Sir Edmund Hillary and Tenzing Norgay told reporter Michael Hellier that, "Exhausted men pitting their strength against a mountain may well see anything. A meeting with dead relatives or friends is typical."

Some of the experienced adventurers disagree, saying they know what a hallucination is and this wasn't it. Their experience is sometimes seen as a psychological compensation that brings inner resources not available in ordinary ways.

Less striking but similar experiences happen to people not under duress. Studies of people who have lost loved ones frequently report that the bereaved persons feel the presence of the deceased loved one, sometimes for years.

Many of these experiences seem spooky, if you'll excuse that characterization. When they're bad things you blame your bad luck. You don't blame a demon who's after you because it sounds crazy and other people tend to stay away from crazy people or people being attacked by demons. However, when good things happen we often credit our good luck and sometimes our "guardian angel" or, if we're especially conceited, that God is looking after us.

Why Things Happen

Let's try to use reasoning rather than mysticism. By understanding the components of living things (physical body, essence, habitat, and culture), you fashion your initiatives to have the best chances for success. The first step is to accept that all of life's events have a probability of happening or not happening, and you can sometimes help to encourage an event with preparations of your own. What you can and can't do is described by two opposing philosophies.

One philosophy, often referred to as *free will* and promoted by motivational speakers and by some religions, is that you're responsible for everything that happens in your life. Where you live, whom you meet, the problems, successes, and failures you have, are all your own doing. If you believe you can do something, you can do it; conversely, if you have negative thoughts/feelings about your chances for success, you won't do it. [Note 3] The idea, of course, is to encourage you (to borrow a recruiting slogan from the U.S. military) "to be the best that you can be." This works well when things go right, as you can imagine. But when things go wrong, there's no one to blame but yourself and any bad feelings you have about your abilities, worth, etc. are magnified.

Another philosophy, the opposite of free will, is called *determinism*. Rather than having complete control and responsibility for your life, determinism posits that every event, decision, and action has been pre-determined by an unbroken chain of events that preceded it. In other words, you have no control over anything and can't change your already-ordained future, so you might as well just watch it happen. As you can imagine, there are many detractors of this philosophy (some of whom call it *fatalism*) because it gives an easy excuse for failure. After all, if I failed, it was because of the unbroken series of previous actions, not because I actually failed. Of course, if I succeeded, it was because I arranged the unbroken series of previous actions to cause my success.

The detractors may scoff, but some varieties of determinism have been shown to have scientific validity. The one most closely aligned with pure determinism is called *causal determinism*, in which past and present events are combined with the laws of nature to predict what will happen next. This works best with physical events such as weather forecasting. It goes like this – if you know the weather we have had and the weather we're having now and the physical laws that govern every aspect of weather (wind patterns, moisture, clouds, sun, etc.) down to the smallest detail, you can predict the future weather. Don't forget flying birds, airplanes, and that annoying volcano that might alter atmospheric patterns. Your prediction may fail because the complex nature of everything involved is too hard to analyze, but theoretically it could be done.

Your Life of Chance

If you're wondering why things happen to your physical body, *biological determinism* has the answer. How your body behaves and reacts to everything in your environment (which includes your habitat and your culture) is completely determined by your genes. Enjoy playing tennis? That's because your genes enjoy playing tennis. Have you decided to go back to school and do research on the essence? Your genes have decided that, not your environment or by reading this riveting book. Some biological determinism adherents also say the religious concept of "soul" is really an expression of your genes and they would say the same about the essence.

Unfortunately, the concepts of biological determinism have also been used by racists to promote "genetic cleansing", saying that certain genes identify criminals or individuals undesirable to those racists. A lot of important work still needs to be done to understand how your genes affect your life, but supporting that work becomes more difficult when it's used for anti-social purposes.

Another variety of determinism, *environmental determinism*, also suffers from misuse. Environmental determinism suggests that the physical environment, rather than social conditions, determines culture. The idea is that the physical environment, especially climate, strongly influences humans to behave in certain ways, and those ways become their culture. For example, the nomadic culture of people living in the desert was caused by the climate creating conditions that didn't permit farming. Unfortunately, this line of reasoning was picked up by colonizers and racists to justify their own superiority (or another's inferiority), and that connection caused a backlash by scientists to reject any connection between environmental factors and culture. Some connections do, of course, exist, such as the presence of a port encouraged a culture of traders in Singapore. While culture and habitat are two of the four components of living things, the connections with the essence are important for your essence. We'll leave for others to do detailed investigations of the factors between culture and habitat and the corresponding influence of environmental determinism.

We can extend the environmental influence on our personal behavior one step further. *Cultural determinism* says that the culture in which we're raised determines our emotions and behavior. Since religion is often a major part of a culture, the effects of our religion on our physical body (behavior) and essence (personality) support the interactions between those living thing components and the culture component.

And, as you may expect, there's *theological determinism* that says a monotheistic God determines everything we do. People who believe in an active, wrathful Supreme Being broaden theological determinism to include divine intention for natural disasters. People often don't understand why a particular natural disaster struck and Omar Sacirbey reports that some religious leaders are ready to tell them:

- Iranian cleric Hojatoleslam Kazem Sedighi told his Shiite Muslim followers that immodestly dressed and promiscuous women are to blame for earthquakes
- Rabbi Yehuda Levin of the Rabbinical Alliance of America warned that "The practice of homosexuality is a spiritual cause of earthquakes"
- Conservative Christian religious broadcaster Pat Robertson blamed Hurricane Katrina on New Orleans' debauchery and immorality
- Israel's Sephardic Chief Rabbi Shlomo Amar said the 2004 South Asian tsunami was "an expression of God's great ire with the world" and Malaysian Muslim cleric Azizan Abdul Razak said it was a message from God
- Conservative Christian commentator Rush Limbaugh said the eruption of a volcano in Iceland was because God was angry about health care legislation recently passed in the United States

Your Essence, Your Eternity

Most religious leaders decry such statements as both false theology and complete nonsense, expressed by people claiming special knowledge of how and why their Supreme Being operates. However, there are plenty of people who want to believe that their views of human activities are also God's views, and God expresses his displeasure when those views are not followed. It's a confirmation of their personal beliefs. As Rabbi Michael Lerner, president of the Network of Spiritual Progressives says, "You start blaming the victims for a process that is a result of something that they had nothing to do with."

You may have noted that in the nature vs. nurture debate, *biological determinism* = *nature* while *cultural determinism* = *nurture*. You can be an adherent of determinism no matter what your developmental beliefs are.

Why do so many people suffer with hunger and disease? Why do bad things happen to good people and good things happen to bad people? These questions are hard to answer with either free will or determinism. Oxford University evolutionary biologist Richard Dawkins says, "the universe we observe has precisely the properties we should expect if there is, at bottom, no design, no purpose, no evil and no good, nothing but blind, pitiless indifference."

The answer suggested by the views in this book goes back to our discussion on probability, that all events in our daily lives are chance events. Certainly, you can increase your chances of doing better by improving your mind and body and avoiding dangerous situations. But where you were born, what accidents befall you, what diseases you encounter, etc. are all governed by chance. Of course, chance is affected by the physical body you inherited, the habitat you live in, and the culture you have. For example, if your physical body has an especially strong immune system, the diseases you encounter by chance will have less of an effect on you.

Naturally, activities you work towards are not chance events. Let's say you have an important job interview, but on the way a traffic accident has occurred that makes you late. The preparation for the interview is not a chance event but the accident is. You may go for your first golf lesson and hit a hole-in-one. Based on that chance event, it would be unwise to quit your job to become a professional golfer.

By recognizing the role of chance, individuals and societies can change the odds to make our lives and the living environment for other species better. We've been doing this for millennia by finding cures for deadly diseases, by realizing the effects of poisons on the environment, and generally by replacing ignorance with understanding. If we continue to use advances in science to benefit the well-being of our planet we'll begin to create our own miracles, ones that will be for all living things and ones that will last.

Chapter 12 – Improved You

Even though you've always been almost perfect ("I thought I was wrong once, but I was mistaken"), you've read this book to eliminate the "almost". Improving yourself means improving your essence.

Summary

As noted in the Preface (you did read the Preface, didn't you?), there's the possibility that your essence may exist forever. We started to better understand our essence by recognizing the roles of the four components of living things and how they interact – your physical body, your essence, your culture, and your habitat. You found out what your essence is and how it works with your body. Your essence is reflected in your personality, and you're trying to improve your personality, but what is your personality? Six of the most influential psychologists in recent history – Gordon Allport, Carl Rogers, Erich Fromm, Abraham Maslow, Carl Jung, and Viktor Frankl – gave you their perspectives on a healthy personality, which were condensed to the single concept followed by virtually every religion – the Golden Rule, "Do to others as you would have others do to you".

Nice, but too theoretical. You need specifics and that's what you got, starting with how personalities develop and mature and then getting personal with your personality by looking at the Big Five Personality Factors of Openness, Conscientiousness, Extroversion, Agreeableness, and Neuroticism, detailed as 26 factors that form the person you are. You recognize the complex interaction of your behavior (part of your physical body and caused by hormones, enzymes, and the various chemicals produced by your body) and your personality (part of your essence), including the problems you have when your personality is in conflict with your body chemicals. And, if you've been blaming some of your habits on bad luck, you found out there's no such thing. So, now what?

Improvement Plan

Now take a cold look at yourself in the mirror and decide what to do. Use an actual mirror – you see more of the real you that way.

Review the numbers you wrote down for yourself for each of the personality attributes in Chapter 9. What, you didn't write them down? Go back and do that.

Improved You

Happy? If so, congratulations! You're as perfect as you want to be. But if you're not happy with some, now's the time to improve them because just getting this far means you have a motivated Achievement Attitude.

Here's another "but". But be sure you're really unhappy and really want to change, because changing your personality takes a lot of dedicated work and you want to be sure the change will make you better, not just a different person. Just becoming a different person with the same issues expressed in a different way makes you a butt-head.

OK, you've decided to make some changes. If you think you're too fat and can't seem to lose weight, keep your mouth shut. If you want to be smarter, turn off the TV and read a book. To make non-physical changes, follow the advice in the "improving" paragraphs for the attributes you want to fix, adjust the advice for your particular life situation, and go for it! But go slowly to see what works and what doesn't work for you.

Perhaps a process used by the National Aeronautics and Space Administration (NASA) to improve software, modified to improve personalities, could help (Figure 12-1).

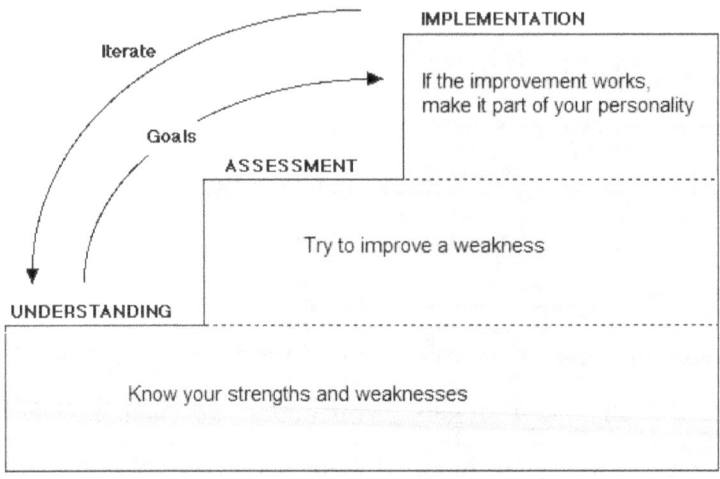

Figure 12-1. Personality Improvement

Here's how this process improvement is done. Look honestly at your personality and see both the good and the bad. Start with one thing you want to improve and try to fix it. If that fix is better, keep being that way and then look at one other thing. Over time this iteration process will make you feel better and better. It makes no sense to go through life feeling unhappy with yourself.

And the changes you make to your personality now will establish the essence you start with in your next physical body. You should appreciate the wonderful opportunity you have of being in a human physical body because other living things have almost no flexibility to change their personalities. Think about it – other animals spend all their time finding food, trying not to be some other animal's next meal, sleeping, and mating. You're at the top of the food chain so you don't have to worry about being eaten, and every variety of our human culture gives you the luxury of a much broader essence.

Some people may think that since it's more likely you'll inhabit a non-human physical body next time it'd be better to increase your viciousness and greed to survive. Sorry – it doesn't work like that. Animal cultures are based on group cooperation and especially vicious or greedy individuals are expelled. If there's one thing you don't want to be in an animal culture, it's expelled. That's a death sentence. In some species the individuals live solitary lives and excessive viciousness and greed may improve your well-being in those cases. But some of those species interact with humans. Do you really want humans to see you as especially vicious? And keep in mind that jails, hospitals, and cemeteries are full of people who have increased their viciousness and greed.

You have this rare opportunity to improve your essence or to make it worse. What you do or don't do is completely up to you. You won't give up because you have a motivated Achievement Attitude, remember? You'll feel positive changes in yourself and you'll see improvements in your relationships with others. As Claude Shannon, who developed the math that makes the Internet possible, said:

Your Essence, Your Eternity

We may have knowledge of the past but cannot control it; we may control the future but have no knowledge of it.

And you know how to develop the best essence, to be the ultimate Improved You, don't you? Follow the Golden Rule.

Notes

1. While the essence is most concerned with personality traits expressed as attributes, a summary of major personality theories may be interesting to readers. Those major theories are *psychoanalytic theories, behaviorist theories, cognitive theories, humanistic theories,* and *biopsychological theories.*

    Psychoanalytic theories suggest that the unconscious mind and childhood experiences form the personality. The founder of these theories was Sigmund Freud who, while trying to understand physical problems ailing people who had no known diseases, discovered that traumas suffered as children were the root cause. By encouraging these people to recall those long-forgotten events, Freud was able to cure their physical problems. While his work mainly applies to mental illnesses and abnormal psychology, it has also been applied to psychological development and personality.

    Behaviorist theories suggest that interaction between the individual and the environment, rather than internal thoughts and feelings, develops the personality. Pavlov's experiments, in which he noticed that dogs salivate when given food, are classics in behaviorist theory. Pavlov rang a bell when he fed his dogs and, after a while, the dogs would salivate when the bell rang, even if no food was given. This *conditioned reflex* created dogs that salivate for no sensible reason, other than they were trained to do so. B.F. Skinner went one step further, training rats that they could obtain food by pressing a lever, creating an association between a behavior and a consequence for that behavior.

Cognitive theories suggest that expectations about other people develop personality. A main proponent was Albert Bandura, who performed a classic exercise called the *Bobo Doll experiment*. He showed a videotape to a class of kindergartners of a college student kicking and verbally abusing a bobo doll. These children then went out to play and saw bobo dolls and nearby hammers. Some of these children began beating the dolls with the hammers, a behavior created by video suggestion alone. The thinking was that the children judged this behavior as appropriate based on the video they observed.

Humanistic theories suggest that free will and individual experience develop personality. In keeping with this approach, psychiatry based on humanistic theory has the therapist act as a facilitator rather than a director, with the client leading the direction. A main proponent of humanistic theories was Carl Rogers, who emphasized compassion and democracy in his dealings with people.

Biopsychological theories suggest that genetics and heredity are responsible for personality. One of the best-known proponents is Hans Eysenck, who believes that introverts avoid stimulation due to high cortical arousal, whereas extraverts seek out stimulating experiences because of low cortical arousal. There have also been studies by Richard Davidson at the University of Wisconsin-Madison on asymmetry in the prefrontal cortex and amygdala affecting personality, especially those with non-verbal learning disorders.

2. How do you define personality traits? Ask ten psychologists and you'll get ten different answers. There have been many attempts over time to understand personality attributes, and many models created to do it.

In 1946 Raymond Cattell used the mathematical technique of factor analysis to develop his wonderful-sounding list of 16 independent personality factors. Those factors are warmth, reasoning, emotional stability, dominance, liveliness, rule-consciousness, social boldness, sensitivity, vigilance, abstractedness, privateness, apprehension, openness to change, self-reliance, perfectionism, and tension. Two problems quickly became evident: 1) Factor analysis only resulted in a few (five or so) of the attributes on the list, although there were plenty of heated arguments about who was doing what wrong. Even Cattell couldn't replicate his original findings. 2) Psychologists preferred attributes that were related rather than independent. They simplified Cattell's list into the five-factor model (Openness, Conscientiousness, Extraversion, Agreeableness, and Neuroticism), which is very popular with psychologists today. Since neither Cattell's list nor the five-factor model is quantitative, I prefer to use the more extensive list by Cattell for essence analysis.

The Zamora Personality Test, developed in 2004 by Antonio Zamora, uses 10 categories of individual attributes and 10 categories of social attributes to gain insights into your personality. Zamora's intention was to find attributes that people find socially desirable, to find a good match for a life partner. As he says, "The test was developed by creating an inventory of characteristics that people wanted in their ideal mate from an extensive compilation of personal advertisements in newspapers." Not exactly rigorous science, but the characteristics he chose do seem to describe personality traits. The individual attributes are achievement attitudes, emotional temperament, energy level, intellectual factors, material attitudes, maturity, philosophical attitudes, physical attributes, risk attitudes, and task performance attitudes. The social attributes are aggressiveness, control attitudes, dependability, egocentrism, emotional expression, fairness, leadership attributes, physical appearance, regard for rules, and team spirit. The Zamora Personality Test focuses on external behavioral traits relating to feelings and beliefs, making it very relevant for essence descriptions.

3. There's a well-known poem by Walter Wintle on positive thinking titled, *Thinking*. The following may be the closest rendition to the original, published in 1905 by Unity Tract Society, Unity School of Christianity:

> If you think you are beaten, you are;
> If you think you dare not, you don't.
> If you like to win, but you think you can't,
> It's almost certain you won't.
>
> If you think you'll lose, you're lost,
> For out of the world we find
> Success begins with a fellow's will;
> It's all in the state of mind.
>
> If you think you are outclassed, you are;
> You've got to think high to rise.
> You've got to be sure of yourself before
> You can ever win a prize.
>
> Life's battles don't always go
> To the stronger or faster man,
> But soon or late the man who wins
> Is the man WHO THINKS HE CAN!

References

Abbott, Derek; Davies, Paul; & Pati, Arun (eds) 2008. *Quantum Aspects of Life*. London: Imperial College Press

Abbott, Edwin A. (1884). *Flatland: A Romance of Many Dimensions*. Amherst New York: Prometheus Books

Albarracin, Dolores; Hart, William; Brechan, Inge; Merrill, Lisa; Eagly, Alice; and Lindberg, Matthew (2009). *Feeling Validated Versus Being Correct: A Meta-Analysis of Selective Exposure to Information*. Psychological Bulletin Vol. 135, No. 4

Allport, Gordon (1955). *Becoming: Basic Considerations for a Psychology of Personality*. New Haven: Yale University Press

Baer, Richard K., M.D. (2007). *Switching Time: A Doctor's Harrowing Story of Treating a Woman with 17 Personalities*. New York: Crown Publishing Group

Benedict, Ruth (1934). *Patterns of Culture*. New York: Riverside Press

Bennett, Craig and Baird, Abigail. *Anatomical Changes in the Emerging Adult Brain: A Voxel-Based Morphometry Study*. In "Human Brain Mapping". Vol. 27, Issue 9, p. 766-777

Biondo, Alessio Emanuele; Pluchino, Alessandro; Rapisarda, Andrea; and Helbing, Dirk (July 2013). *Are Random Trading Strategies More Successful than Technical Ones?* Paper submitted at Cornell University

Bouchard, Thomas J, Jr.; Lykken, David T.; McGue, Matthew; Segal, Nancy L.; and Tellegen, Auke (October 12, 1990). *Sources of Human Psychological Differences: The Minnesota Study of Twins Reared Apart*. Science, New Series, Vol. 250, No. 4978, p. 223-228.

Bryson, Bill (2003). *A Short History of Nearly Everything.* New York: Broadway Books

Byrnes, James; Miller, David; & Schafer, William. (1999). *Gender Differences in Risk Taking: A Meta-Analysis.* Psychological Bulletin, 125: 367–383

Carmody, Dennis; Dunn, Stanley; Boddie-Willis, Akiza; DeMarco, J. Kevin; and Lewis, Michael (September 2004). *A Quantitative Measure of Myelination Development in Infants, Using MR Images.* Neuroradiology 46(9), p. 781-786

Carnegie, Dale (1936). *How to Win Friends and Influence People.* New York: Simon & Schuster

Cattell, Raymond. B. (1990). *Advances in Cattellian Personality Theory.* In L. A. Pervin (Ed.), "Handbook of Personality: Theory and Research" (pp. 101-110). New York: Guildford.

Davidson, Richard et al. (2003). *Alterations in Brain and Immune Function Produced by Mindfulness Meditation.* Psychosomatic Medicine 65, p. 564-570

Dawkins, Richard (1995). *River Out of Eden: A Darwinian View of Life.* New York: BasicBook

Dennett, Daniel (1991). *Consciousness Explained.* New York: Back Bay Books

DeBruine, Lisa (2002). *Facial Resemblance Enhances Trust.* Proceedings of the Royal Society of London B, 269(1498). p. 1307-1312

Deci, El. L. and Ryan, R. M. (2000). *The "What" and "Why" of Goal Pursuits: Human Needs and the Self-Determination of Behavior.* Psychological Inquiry 11: p. 227-268

DeYoung, Colin (2010). *Brain Structure Corresponds to Personality.* Psychological Science Vol. 21 No. 6

References

Eddington, Neil and Shuman, Richard (2006). *Healthy Personality.* Presentation given by Continuing Psychology Education

Einstein, Albert (1905a). "Über einen die Erzeugung und Verwandlung des Lichtes betreffenden heuristischen Gesichtspunkt". *Annalen der Physik* **17** (6): 132–148

Einstein, Albert (1905b) "Zur Elektrodynamik bewegter Körper", *Annalen der Physik* 17: 891

Elbert, Thomas; Pantev, Christo; Wienbruch, Christian; Rockstroh, Brigitte; and Taub, Edward (October 13, 1995). *Increased Cortical Representation of the Fingers of the Left Hand in String Players.* Science, Vol. 270, p. 305-306

Emmons, Robert A. (2007). *Thanks! How the New Science of Gratitude Can Make You Happier.* Boston: Houghton Mifflin.

Finkel, Michael (December 2009). *The Hadza.* National Geographic Vol. 216, No. 6, p. 94-119

Frankl, Viktor (1962). *Man's Search for Meaning: An Introduction to Logotherapy.* Boston: Beacon Press

Freund, Julia; Brandmaier, Andreas M.; Lewejohann, Lars; Kirste, Imke; Kritzler, Mareike; Krüger, Antonio; Sachser, Norbert; Lindenberger, Ulman; and Kempermann, Gerd (May 10, 2013). *Emergence of Individuality in Genetically Identical Mice.* Science Vol. 340 No. 6133, p. 756-759

Fromm, Erich (1955). *The Sane Society.* New York: Holt, Rinehart & Winston

Fuster, Joaquín (2008). *The Prefrontal Cortex.* Amsterdam: Academic Press

Gallozzi, Chick (2010). *Spiritual Consciousness.* www.personal-development.com/chuck/spiritual-consciousness.htm

Gaylor, Thomas Kent (2001). *Factors Affecting Resistance to Change: A Case Study of Two North Texas Police Departments.* Master of Arts Thesis: University of North Texas

Ger, Güliz and Belk, Russell (1999). *Accounting for Materialism in Four Cultures.* Journal of Material Culture, Vol. 4, No. 2, p. 183-204

Giedd, Jay (2008). *Inside the Teenage Brain.* Interview with the Public Broadcasting System

Gräff, Johannes; Joseph, Nadine F.; Horn, Meryl E.; Samiei, Alireza; Meng, Jia; Seo, Jinsoo; Rei, Damien; Bero, Adam W.; Phan, Trongha X.; Wagner, Florence; Holson, Edward; Xu, Jinbin; Sun, Jianjun; Neve, Rachael L.; Mach, Robert H.; Haggarty, Stephen J.; Tsai, Li-Huei (January 16, 2014). *Epigenetic Priming of Memory Updating during Reconsolidation to Attenuate Remote Fear Memories.* Cell, Vol. 156, Issue 1, p261–276

Gray, John (1992). *Men Are From Mars, Women Are From Venus.* New York: HarperCollins

Haidt, Jonathan and Joseph, Craig (Fall 2004). *Intuitive Ethics: How Innately Prepared Intuitions Generate Culturally Variable Virtues.* Daedalus, American Academy of Arts & Sciences, p. 55-66

Harlow, Harry (December 1958). *The Nature of Love.* American Psychologist. Vol. 13 (12), p. 673-685

Harrell, Eben (November 28, 2011). *A Flicker of Consciousness.* Time, Vol. 178, No. 21, p. 42-47

Hauser, Marc (September 2009). *The Mind.* Scientific American vol. 301 no. 3, p. 44-51

Hawking, Stephen (1988). *A Brief History of Time.* Toronto: Bantam Books, p. 140-141

Hawking, Stephen and Mlodinow, Leonard (2010). *The Grand Design.* Bantam Books

References

Hellier, Michael (January 18, 1971). *Mountain Ghosts.* San Francisco Chronicle Huttenlocher, Peter (1990). *Morphometric Study of Human Cerebral Cortex Development.* Neuropsychologia 28, p. 517-527

James, William (1898). *Human Immortality: Two Supposed Objections to the Doctrine* 2nd ed. Boston: Houghton Mifflin

Jochle, Wolfgang (1998). *Fehlverhalten und Anpassungsprobleme bei Hund und Katze und deren pharmakologische Beeinflussbarkeit.* [Abnormal behavior and adaptation problems in dogs and cats and their pharmacologic control]. Tierärztliche Praxis. Ausgabe K, Kleintiere/Heimtiere 26(6): 410-421

Johnson, Wendy; Krueger, Robert F.; Bouchard, Thomas J., Jr.; and McGue, Matt (April 2002). *The Personalities of Twins: Just Ordinary Folks.* Twin Research

Jung, Carl (1953). *Two Essays on Analytical Psychology.* New York: Pantheon

Klosko, George (2005). *Political Obligations.* Oxford University Press

Kuschel, Rolf (1989). *Vengeance Is Their Reply: Blood Feuds and Homicides on Bellona Island, Part 1: Conditions Underlying Generations of Bloodshed.* Copenhagen: Dansk Psykologisk Forlag

Lange, Karen (January 2002). *Wolf to Woof.* National Geographic, Vol 201 No. 1, p. 2-11

Larsson, Mats; Pedersen, Nancy; and Stattin, Håkan (May 2007). *Associations Between Iris Characteristics and Personality in Adulthood.* Biological Psychology, Vol. 75, Issue 2, p. 165-175

Lewin, Kurt; Lippitt, R; White, R.K. (1939). *Patterns of Aggressive Behavior in Experimentally Created Social Climates.* Journal of Social Psychology 10: 271–301

Maslow, Abraham (1943). *A Theory of Human Motivation.* Psychological Review (50) p. 370-396

McGrath, Susan (March 2014). *Call of the Bloom.* National Geographic Vol. 225 No. 3, p. 128-139

Meussling, Vonne (April 1986). *Attitude: A Component of Competent Performance.* Paper presented at the Annual Meeting of the Central States Speech Association (Cincinnati, OH, April 17-19, 1986)

Milgram, Stanley (1963). *Behavioral Study of Obedience.* Journal of Abnormal and Social Psychology, vol. 67, p. 371-378

Miller, Peter (January 2012). *A Thing or Two About Twins.* National Geographic Vol. 221 No. 1, p. 38-65

Moyer, Michael (September 2013). *The Food Issue.* Scientific American Vol. 309 No. 3, p. 34-39

Newberg, Andrew; D'Aquili, Eugene; & Rause, Vince (2001). *Why God Won't Go Away.* New York: Ballantine Books

Orshansky, Mollie (1963). *Children of the Poor.* Social Security Bulletin: Social Security Administration

Overstreet, Harry A. (1949). *The Mature Mind.* New York: Norton & Co

Pert, Candace (2000). *Everything You Need to Know to Feel Go(o)d.* New York: Hay House

Peterson, Christopher & Seligman, Martin (2004). *Character Strengths and Virtues.* Oxford: Oxford University Press

Piaget, Jean (1932). *The Moral Judgment of the Child.* London: Kegan Paul, Trench, Trubner and Co

Pinker, Steven (1997). *How the Mind Works.* New York: W. W. Norton & Company. p. 374

Ratliff, Evan (March 2011). *Taming The Wild.* National Geographic, Vol. 219, No. 3, p. 34-59

References

Raichle, Marcus (March 2010). *The Brain's Dark Energy.* Scientific American Vol. 302 No. 3, p. 44-49

Rauscher, Elizabeth and Targ, Russell (2001). *The Speed of Thought: Investigation of a Complex Space-Time Metric to Describe Psychic Phenomena.* Journal of Scientific Exploration 15(3): p. 331-54

Rogers, Carl (1961). *On Becoming a Person: A Therapist's View of Personality.* Boston: Houghton Mifflin

Sacirbey, Omar (May 1, 2010). *Fundamental Cause & Effect.* Religion News Service, reported in the Washington Post

Schoonover, Carl (March 2011). *Signals in a Storm.* Scientific American Vol. 304, No. 3, p. 46-47

Seaward, Brian L. (1995). *Reflections on Human Spirituality for the Worksite.* American Journal of Health Promotion 9, No. 3: p. 165-168

Seyle, Hans (1956). *A Syndrome Produced by Diverse Nocuous Agents.* Nature

Smith, Adam (1776). *Wealth of Nations.*

Somé, Malidoma (1998). *The Healing Wisdom of Africa: Finding Life Purpose Through Nature, Ritual, and Community.* New York: Jeremy P. Tarcher/Putnam. p. 31 & 38

Sponsel, Leslie E. (1996). *The Natural History of Peace: A Positive View of Human Nature and Its Potential* in "A Natural History of Peace", edited by Thomas Gregor, p. 95-125. Nashville, TN: Vanderbilt University Press

Stafford, Candice A.; Walker, Gregory P.; and Ullman, Diane E. (May 23, 2011). *Infection With a Plant Virus Modifies Vector Feeding Behavior.* Proceedings of the National Academy of Sciences

Stevenson, Ian (1997) *Where Reincarnation and Biology Intersect.* Westport, Connecticut: Praeger

Suedfeld, Peter and Geiger, John (2008). *The Sensed Presence as a Coping Resource in Extreme Environments* in "Miracles: God, Science, and Psychology in the Paranormal", edited by J. Harold Ellens, Vol. III. Westport, Conn.: Praeger

Surowiecki, James (2004). *The Wisdom of Crowds*. New York: Doubleday

Thomas, Elizabeth Marshall (1958). *The Harmless People*. New York: Vintage Books

Tiller, William (2008). *Toward a Reliable Bridge of Understanding Between Traditional Science and Spiritual Science* in "Measuring the Immeasurable". Boulder CO: Sounds True. p. 287-312

Toffler, Alvin (1970). *Future Shock*. New York: Bantam Books

Trompenaars, Fons (2003). *Did The Pedestrian Die?* Chichester, England: Capstone Publishing Limited

Tucker, Jim B (2005). *Life Before Life*. New York: St. Martin's Press

Tyler, Anne. (1985). *The Accidental Tourist*, New York: Berkley Books

Vyse, Stuart. (1997). *Believing in Magic: The Psychology of Superstition*. New York: Oxford University Press

Warber, Sara and Irvine, Katherine (2008). *Nature and Spirit* in "Measuring the Immeasurable". Boulder CO: Sounds True, p. 137

Watson, James (1968). *The Double Helix*. New York: Antheneum

Weber, Elke and Hsee, Christopher (1999). *Models and Mosaics: Investigating Cross-Cultural Differences in Risk Perception and Risk Preference.* Psychonomic Bulletin & Review, Vol 6, No. 4, p. 611-617

Williams, Lawrence E. and Bargh, John A. (October 24, 2008). *Experiencing Physical Warmth Promotes*

References

Interpersonal Warmth. Science Vol. 322 No. 5901: 606-607

Williams, Leanne; Brown, Kerri; Palmer, Donna; Liddell, Belinda; Kemp, Andrew; Olivieri, Gloria; Peduto, Anthony; and Gordon, Evian. *The Mellow Years?: Neural Basis of Improving Emotional Stability over Age.* Journal of Neuroscience, June 14, 2006, 26(24):6422-6430

Witten, Edward (1995). "String theory dynamics in various dimensions". *Nuclear Physics B* **443** (1): 85–126.

Wynne-Edwards, Kathryn E. (September 2001). *Hormonal Changes in Mammalian Fathers.* Hormones and Behavior, Volume 40, Number 2, p. 139-145(7)

Yuste, Rafael and Church, George M. (March 2014). *The New Century of the Brain.* Scientific American Vol. 310, No. 3, p. 38-45

Zimmer, Carl (January 2011). *100 Trillion Connections.* Scientific American, Vol. 304 No. 1, p. 58-63

Your Essence, Your Eternity

About the Author

Howard Jeffrey Bender has had a long career both in and supporting scientific research. At Penn State University he helped study Jupiter's Great Red Spot using radio astronomy and discovered a new variety of tin while assisting in high-pressure physics research. He was a computer scientist at NASA for 13 years and taught software engineering at the University of Maryland for 27 years. He has written papers on applications of String Theory and DNA/RNA research and has assisted in environmental biodiversity research in the U.S., Japan, France, Costa Rica, Cayman Islands, and the Galapagos. He was recognized for his visual improvement system by a Johns Hopkins University competition for *Personal Computers to Aid the Handicapped* and was awarded a patent for a process computers use to understand human languages. He's a mediocre bridge player, a very mediocre tennis player, and an extremely mediocre banjo player, known for being able to clear a room in less than a minute. He has a B.S. from Penn State University, an M.S. from the Polytechnic Institute of New York, and a Ph.D. from the University of Maryland.

www.ingramcontent.com/pod-product-compliance
Lightning Source LLC
Chambersburg PA
CBHW060820220526
45466CB00003B/922